A Field Guide to Time-Varying Light Sources

Seeing in the Fourth Dimension

Daniel H. Bennett

A Bear Peering Round A Rock

A Field Guide to Time-Varying Light Sources

Seeing in the Fourth Dimension

Daniel H. Bennett

Cover by Daniel H. Bennett

Library of Congress Control Number: 2015900348

ISBN-13: 978-069-236090-3

ISBN-10: 0-692-36090-5

Website and On-Line Forum

timevaryinglights.com

Cover image: A swept photographic trace of neon tube decoration on Kellys Brew Pub in Albuquerque, New Mexico. The camera was rotated and the lens zoomed during the 1/10 second shot.

Published through CreateSpace – www.createspace.com

A Bear Peering Round A Rock

Fort Collins, Colorado, USA peeringbear.com

About the author

Photograph © 2012 Hyeon-Na Woo. Used by permission.

Daniel H. Bennett was born in London in 1965 and lived in England until moving to the United States in 2000.

A degree in electronic and electrical engineering took him into a career that's been focused on high end desktop computers and displays, with expertise in computer graphics and display technologies. Among his passions is to communicate how technology works to audiences at any level, from school children to experienced professionals.

His interest in vision and observation of the behavior of light has been life-long. On a trip to France as a young man his hosts called him "Monsieur Tête-en-l'air" (Mr. Head-in-the-air) due to his always watching for halos, rainbows, sun-dogs (parhelia), etc. He was particularly influenced by a book written in 1939 (published in English in 1954, Dover) by Professor M. Minnaert, called *The Nature of Light and Colour in the Open Air*. The book you are holding was conceived as a modern successor to that book, and adopts a similar approach, while revealing the secrets of modern light sources.

Daniel is married with four children, a motorcycle and two accordions.

Dedication

To my parents, for buying that oil-filled radiator

Contents

Introduction

What you'll get out of this book: a whole new way of seeing

Welcome to a unique book and a unique way of looking at the world.

This book will show you that by learning one or two simple techniques you will be able to see – with your own eyes – what you probably never knew was there: the *time dimension* of many sources of illumination.

Although there have been man-made light sources for millennia, the last thirty or forty years have seen radical technological changes. Gone are candles, tallow lamps, burning torches and storm lanterns; gas lamps are a thing of the past, and even Thomas Edison's ubiquitous incandescent light bulb is being challenged by new forms of lighting.

One characteristic of many modern light sources is that although they appear to shine with a steady light, they are in fact flashing or modulating very rapidly – too rapidly, in fact, for the human eye to see. This is by design: they're *supposed* to look like steady light sources. However, by applying the techniques detailed in this book, you'll be able to unravel every nuance of the behavior of these light sources, no matter how quickly they flash or change.

Figure 1: All the equipment you'll need: a small mirror.

A simple pocket mirror is the only tool you'll need to open up the world of time-varying light sources. And if you're really good, you don't even need the mirror!

In addition, there is a chapter on photography, which explains how to obtain photographs like the ones throughout this book; and you'll see that with simple techniques and a little imagination it's possible to produce striking images and photographic art.

The bulk of this book is the Field Guide itself, which arranges time-varying light sources into categories and presents common examples of each type of light source. Photographs show you what you might see, and observation tips (indicated by this symbol:) help you to look for important features.

Quantum physics and relativity

Relax! We're not going to dig deep into physics (or ocular physiology, for that matter) in this book. Explanations of how light sources behave the way they do will be high-level and simple to grasp, and will be sufficient to explain what you see. Of course, underlying the behavior of all light sources is quantum physics – after all, the photon, which is a particle of light, is described as the *quantum* of electromagnetic radiation, and this book is arguably all about photons; but it's more about *enjoying* them than studying the deepest secrets of their behavior.

And don't worry about Einstein's theories of relativity either! They don't really intersect with this book (except perhaps in the section on pulsars on page 144); except to say that the concept of "space-time", which is a consequence of Einstein's theory of special relativity, regards space has having *four* dimensions, with time being the fourth one. Hence the sub-title of my book: *Seeing in the Fourth Dimension.*

How I came to write this book

When I was between two and four years old, I lived with my parents and younger brother in the pretty Suffolk town of Woodbridge in England. I remember very little about that time in my life, but two memories stand out.

Firstly, I discovered ocular parallax. Close one eye and then the other and everything jumps sideways. (Yes, I know there's more to it than that, but it's outside the scope of this book.)

And secondly (and here's where this book started), I discovered how to view time-varying light sources. In my bedroom was a free-standing oil-filled electric radiator. It had an orange-red neon indicator which showed when the radiator was switched on. At night, that little red light was the only light source in my room, and I noticed that if I jiggled my gaze around the room, I would see a scribbled line of red dashes. Of course, at the tender age of four I didn't understand about such things as AC electricity, neon indicator lamps and retinal persistence of vision; but the interest was born.

Figure 2: A radiator similar to the one that was in my bedroom. Image © Glen Dimplex Group. Used by permission.

Now, more than forty-five years later, I've enjoyed many years of delighted and inquisitive observation of time-varying light sources, and it's time to share my experiences and to provide my readers with the keys to the time-varying optical world.

Enjoy the journey!

The web-site

There is a web-site associated with this book: *timevaryinglights.com*. Make sure you go there for updates, new photographs, tutorial videos and a discussion forum.

Time-varying light sources: what they are and where to find them

What is a time-varying light source?

A quick look at the pages of this book will show you that many – most, in fact – of the sources of light that you see at night or in your home and office are time-varying. What does that mean?

Simply put, "time-varying" means that something changes as time goes by.

Some light sources vary slowly with time: imagine a flashing red light at a railroad crossing, for example. The light is certainly time-varying; but it's designed to be that way: you're supposed to notice that it's flashing. Such light sources are not the subject of my book: they're not nearly interesting enough!

More interesting are the sources of light which are flashing – or varying in some other way – too fast for the human eye to see under normal circumstances. These lights are designed to appear steady, and are flashing for reasons of technology or physics. You'll find that a large proportion of lights, lamps and displays fall into this category: these are what this book is about.

Take a very common type of lamp as an example: the traditional fluorescent tube light, seen in garages everywhere. When you switch it on, it may flash once or twice, but then it delivers steady white light. Or so it would seem. Did you know that for about fifty percent of the time, the light is a bright bluish white, and that for the remaining fifty percent, it's much dimmer and a strong yellow or even brown color? You don't ordinarily see this behavior because it happens so fast that the eye's natural "persistence of vision" masks the variations. But those variations are there, and it's easy to see them if you know how.

Other light sources are time-varying in much more complex ways, with the archetype being, perhaps, the old-style cathode ray tube television. That's the old type that had a slightly domed screen and took several seconds to warm up. The early designers of television owe a lot to persistence of vision, as you'll find out on page 81.

Where to find time-varying light sources

Since many man-made light sources these days are time-varying, it's not hard to find them. Go outside at night: street lights, stop lights, car tail-lights, signs in shop windows, even those lights that cyclists strap to their heads – many of them are time-varying. And when you get back to your living room, look at the numeric display on your DVD player, the monitor on your computer and the neon indicator on a power strip. Time-varying, all of them.

Unless you're living in a candle-lit, log-fire-heated cabin in remotest Montana, you're going to find time-varying light sources all around you. This book will help you to see them for what they are.

Figure 3: A vertically swept photograph of the distant street lights of a city. The dashed lines indicate that almost all of the light sources in this scene are time-varying. This illustrates that time-varying light sources are almost everywhere.

Figure 4: High Rye Cabin in Montana's Beaverhead-Deerlodge National Forest. Photograph © 2011 Nathaniel C. Liedl. Used by permission.

How to observe time-varying light sources

Persistence of vision: your friend any way you look at it

All the light sources described in this book have characteristics, such as intensity, color or pattern, which change as time advances. What's more, these variations occur too fast for the human eye to see under normal conditions. This is because the retina, which is the surface in the eye that captures what you see, has a characteristic called "persistence of vision". For a rapidly varying light source where one phase is brighter than another, the retina responds most strongly to the brightest phase of the light's output, and other phase is not perceived because the retina's cells don't "reset" in time to see it before another bright phase occurs.

Let me illustrate this with an example: A typical fluorescent tube light (the kind that flashes a couple of times when you switch it on, and often buzzes) exhibits a cycle something like this:

- Fifty to sixty percent of the time, it emits a bright, slightly bluish white.
- Forty to fifty percent of the time, it emits a much dimmer, yellow or brown light.

The reasons for this behavior are given in the section on this type of lighting, on page 70.

This cycle repeats 120 times per second (in the USA; 100 times per second in Europe).

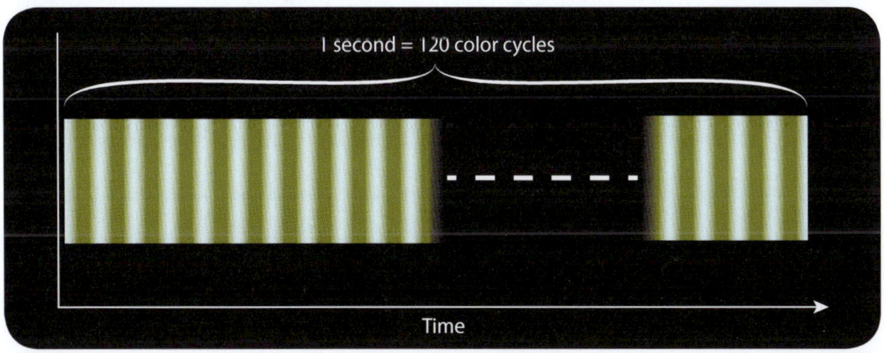

Figure 5: A diagram representing the changing color of a fluorescent tube light on a 60Hz electrical supply.

When you look at a fluorescent tube, the retina responds primarily to the bright white light, and you will never see the yellow phase because the retina's cells don't relax after the bright white illumination in time to see the much dimmer yellow light. After all, the yellow period lasts probably one two-hundredth of a second, which is much too short a time for the retina to respond after the bright period. Then the next bright period comes along and reinforces the perceived bright white color. (Note, however, that the eye does "mix" the colors together: without the yellow phase, the bluish tint of the bright white phase may dominate. The yellow phase might have the effect of making the white a little "warmer".)

Normally, this is desirable: when you light your room with a fluorescent light, you don't want to see everything flashing white and yellow. The designers of fluorescent lighting took advantage of the eye's behavior in order to deliver the apparently steady illumination that you expect from a light in your room. (But ponder for moment: lit by this type of fluorescent light, the illumination in the room really *is* changing rapidly between white and yellow! See Figure 127 if you don't believe me.)

However, if you're interested in observing the time-varying behavior of light sources like a fluorescent tube, you can play some tricks which turn the retina's persistence of vision to your advantage.

What I wrote above, about the bright white phase overriding the dim yellow phase is only true *if the image of the fluorescent tube remains in one place on your retina.* In other words, as long as you look straight at the tube, you will never see the yellow phase.

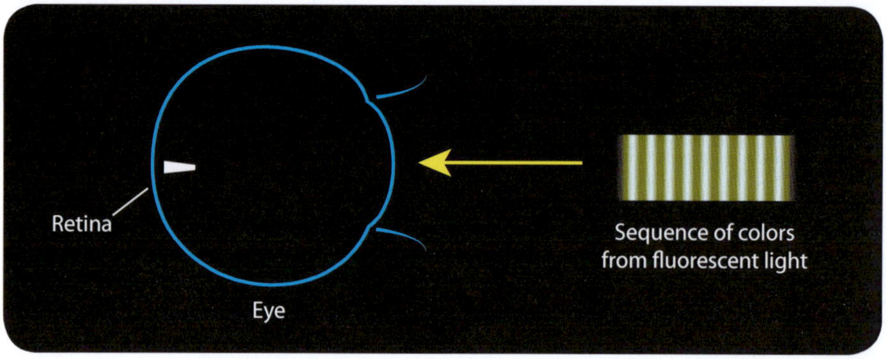

Figure 6: If the image of a fluorescent tube remains in one location on the retina, the sequence of colors will not be seen. Instead, the colors will be combined into the perception of a steady white image.

However, if you can somehow present fresh areas of the retina – which have not been exposed to the bright white phase – to the image, you will see the dimmer yellow phase very clearly, albeit briefly.

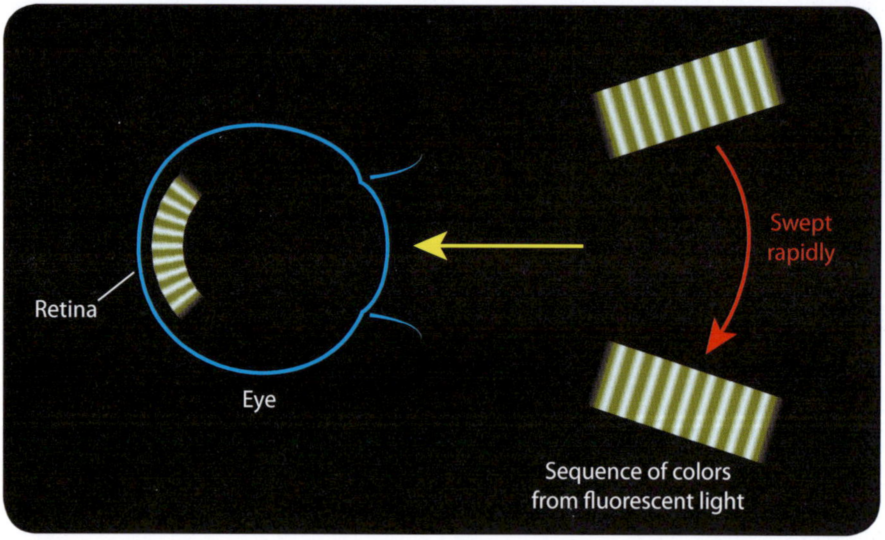

Figure 7: In this case, the eye's direction of view is swept – or the fluorescent tube is moved, or a mirror is used – so that its image is swept rapidly across the retina. The result is a short-lived (I use the word "transient") image on the retina which "unpacks" the color variations. This allows the observer to see exactly how the light source is behaving over time.

And that is the basis of how to see time-varying light sources: *keep the image moving!*

So the question becomes: how can one keep the image moving?

One of the easiest ways to see the rapid variations of a light source's brightness is to watch while somebody moves the light source around: try it in a dark room with an electrical power strip that has a red neon indicator on it.

However, in most cases you won't be able to move the light source (try waving a row of street lights around!), so you will need to find another way to make the image of the light source move across your retinas. The following sections detail how this can be done. You should also watch the tutorial videos on the "How To See" page of the web-site, timevaryinglights.com

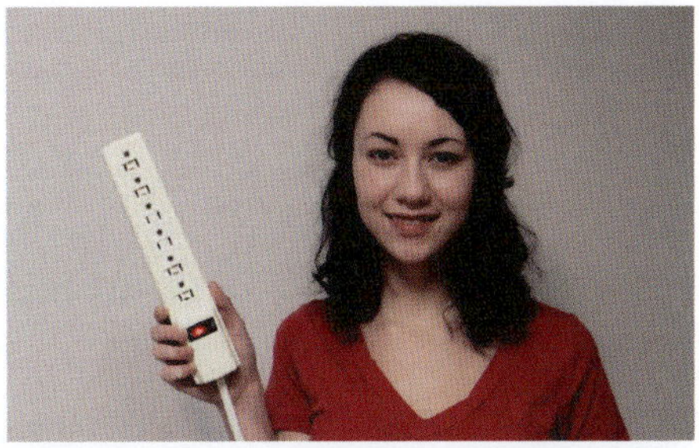

Figure 8: One of the easiest ways to see a time-varying light source is to have someone wave an electrical power strip in front of you, especially in low light conditions. Note the red or orange neon indicator: this is the light source.

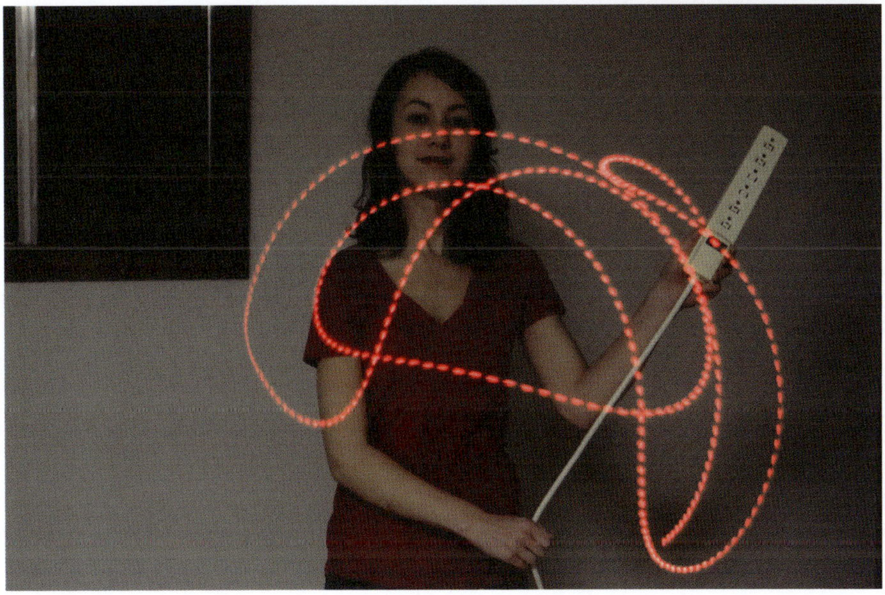

Figure 9: This photograph was created using a 2.5 second exposure, during which time my daughter waved the power strip in wide sweeps in a dark room. The dashed nature of the trail is due to the fact that the neon light is flashing on and off rapidly. If the neon light were steadily illuminated, the trail would be a solid, undashed line. For interested photographers: I set the flash to fire at the start of the exposure, to capture the scene; but I reduced the flash intensity (using the flash compensation setting on the camera) to under-illuminate the scene so that the trail of dashes from the neon lamp would stand out.

Vision jiggling

"Vision jiggling" is a term I will use throughout this book, and it refers to any technique you may use to move the image of a light source rapidly across your retinas, to achieve the kind of observations I describe. For this book, there are two main methods for vision jiggling: *eyeball jiggling* and *mirror jiggling*. Both will work; and they are described below.

Figure 10: The MOTORPLEX sign referred to in the text. All letters are time-varying, except for the "M", for reasons known only to the sign-maker. (1/10 sec)

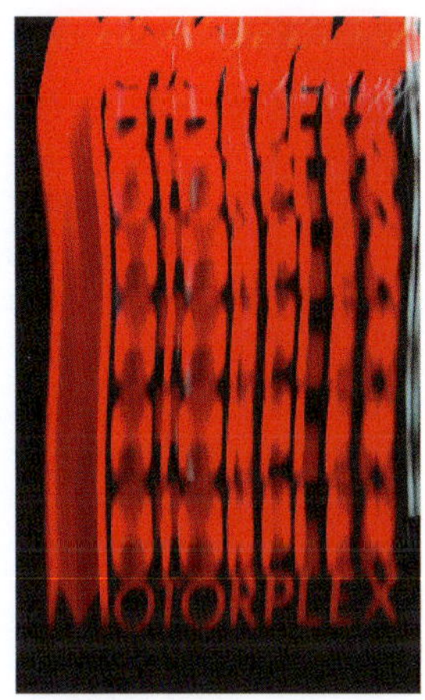

Eyeball jiggling

Eyeball jiggling is something I've been able to do for as long as I can remember.

It is simply the action of repeatedly moving your gaze very rapidly up and down (or side to side) without closing your eyes.

The aim of eyeball jiggling is to keep the image of the light source moving across your retinas, which gives you a "retinal trace" (see page 17). Eyeball jiggling is the best technique to use because it doesn't require any equipment, such as a mirror. I can use it while driving at night, for example. With practice, a quick flick of the eyes is all it takes to reveal the time-varying behavior of a light source.

In fact, being able to eyeball-jiggle lets you see time-varying light sources that you weren't even looking for. For example, once when I was driving I saw a large illuminated sign that said "MOTORPLEX", and I noticed immediately that while all the letters appeared to be the same red color, the "M" was not time-varying but the other letters were (probably lit with fluorescent tubes). It's the joy of being able to "just see things" like that which makes it worthwhile to learn to eyeball-jiggle.

Figure 11: Move your gaze up-down-up-down repeatedly, as smoothly and rapidly as possible, without closing your eyes.

I do understand, however, that the muscular skill needed for eyeball jiggling doesn't come immediately to everyone. It's the same thing as watching someone play five consecutive notes with five fingers on a piano if they have never done it before: the nerves and muscles simply don't operate that way. But then, that's why we practice.

The type of motion that you should aim for is both rapid and smooth. Rapid jiggling causes the details on the retinal trace to be well-separated, which makes it easier to interpret the light source's behavior. And smooth oscillation means that the trace will be almost linear, allowing you to judge things like the relative time spent in light and dark states during the light's repeated cycle. An uneven motion of the eyes will result in a trace that's bunched and irregular, which is harder to interpret. As you look through the Field Guide section of this book you will see that many of the photographs were captured using a single smooth sweep of the camera, to make the trace as easy to interpret as possible. You should aim for this type of motion. However, repeated oscillation is useful because the trace does not last very long, so a repeated motion allows you to see it several times, which gives you more opportunity to interpret it.

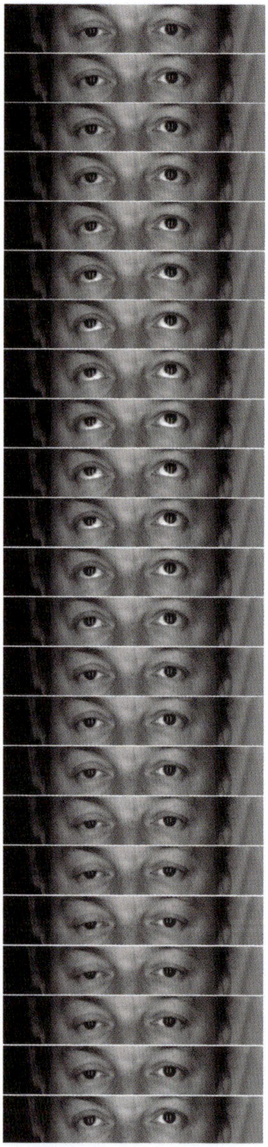

Figure 12: A series of frames from a high-speed video of the author eyeball-jiggling. These frames are 1/112 of a second apart. Courtesy of PhaseSpace Inc.

I have endeavored to train people to jiggle their eyeballs, with mixed success. In a dark room with a clear time-varying light source, it's almost always possible for a beginner to see the retinal trace, at least unintentionally such as when looking from one part of the room to another (a valid technique, by the way). However, being able to oscillate one's eyes does not come easily to most people, so when training I usually concentrate on using a mirror (see the next section) so that my students get to see what they're looking for (retinal traces), and then I encourage them to practice eyeball jiggling over the next few weeks.

Some people are an exception: One friend of mine has for many years been able to move his eyes in a rapid circular motion! However, it wasn't until I alerted him to retinal traces that he was suddenly able to observe time-varying light sources. It was like I opened up a whole world of observation for him! Of course, the fact that we were in Las Vegas at the time made it all the more rewarding. See page 127 for a time-varying portrait of that city.

The fact remains, however, that many people will find eyeball jiggling challenging. For those who want to observe time-varying light sources without the pain of learning to jiggle their eyeballs, there's a simple remedy: use a mirror.

Using a pocket mirror: mirror jiggling

A mirror lets you make all the observations that can be made by eyeball jiggling; however, it's less convenient because you need to carry the mirror around with you. For example, imagine how you can impress your friends by simply glancing (and eyeball jiggling for a second) at a pair of street lights and casually telling them that the lights are wired to different phases of the electrical supply! (OK, so it takes a certain kind of friend to be impressed by something like that, but still...). Now imagine how much less impressive it would be if you have to hunch over a mirror to make your observation!

Therefore, I recommend that if you can train yourself to jiggle your eyeballs, it's worth it.

In the meantime, here is the technique for using the mirror:

Your objective is to cause the image of the light source to move rapidly and smoothly across your retina. This gives rise to a "retinal trace", about which you will learn in the next chapter.

Hold the mirror close to your face, positioned near your cheek, with the mirror tilted about 45° away from you. Look down into the mirror so that you can see the light source you're interested in.

Figure 13: Hold the mirror below your eye and tilt it away from you. When you look down into it, you should direct the mirror so that you can see the light source you wish to observe.

Now rock the mirror back and forth rapidly to make the image travel across your retina. You'll find it easiest if the mirror is quite close to your eye. For example, if you're looking down into the mirror, you might rest your thumb on your cheek bone.

Figure 14: Rock the mirror rapidly. In this sequence of three images, you can see that the angle at which the mirror is tilted is different in each image, to illustrate how you would rock the mirror.

Some people find it quite hard to use their fingers to rock the mirror rapidly back and forth. If you're among them, you may get better results by holding each end of the mirror with fingers of each hand, and using your arms to move your hands and hence the mirror.

Note that – just like with eyeball jiggling – a rapid motion will cause the features in the retinal trace to separate out clearly, making it much easier to observe the light source's time-varying behavior. Therefore, the ability to rock the mirror rapidly is to be cultivated. Additionally, a smooth motion is preferred: if your mirror moves in jerks, the trace will be bunched and irregular, making it much harder to interpret.

At a pinch, the display of a cellphone makes a reasonably good mirror. Sometimes, though, it may have a coating which degrades the image, or the touch sensor may throw colored patterns around the image of the light. But for most purposes, your cellphone may be your handiest mirror.

Figure 15: This photograph was taken with the camera pointing downwards at the mirror. The mirror is tilted away from the camera so that we can see the target light source. In this case, it's a battery charger that's discussed in detail on page 26. Although hard to see in this flash-illuminated photo, there are two red LEDs which are the light sources we are interested in.

Figure 16: The same scene, shot without flash, and with the camera set to 1/10 of a second exposure time. The mirror was tilted rapidly and smoothly through a small angle *while* the photo was taken. This results in the images of the two LEDs being swept across the photo, exposing their time-varying traces.

Watching car roofs: taking advantage of motion

If you have not yet got the knack for eyeball or mirror jiggling, but you would like to see those traces I keep writing about, here's another technique.

Place yourself at night at a little distance from a fairly busy road, in such a location that there are streetlights visible in the middle distance beyond the road. In other words, you should be able to look *across* the road at the streetlights. In addition, select your position such that the lights are visible a little above the roofs of cars that go past you on the road.

The technique is simple: Watch the roofs of cars as they pass. The smooth, rapid motion of a car across your view causes your eyes to move in a perfect sweep. Look for the trails left by the streetlights across your retinas: they will almost certainly be dashed, showing the time-varying behavior of the lights very clearly. Of course, the trails will be horizontal because the motion of your eyes was horizontal as they followed the car.

This technique can't be applied everywhere, of course; but it should serve as an easy way to see the traces which concern this book.

You can take advantage of other movements too. Wherever you can watch something that's moving rapidly against a background that contains time-varying light sources, you should be able to see traces.

Eating and singing

In some cases it's possible to see interesting effects when looking at time-varying light sources while eating, especially if you're eating something crunchy. The vibrations in your head cause momentary disturbances in the direction in which your eyes are pointing, and this can cause the image of the light source to be moved randomly across your retinas. The best kinds of light sources for observing this effect are LED Christmas lights (say on a Christmas tree), or LED numeric displays such as on a digital alarm clock. When you're eating, the lights or digits appear to dance around, appearing to move separately from whatever they're attached to. Turning the room lights off can help you to see this effect.

Singing – at least for men – is another way to see this effect. The technique is to sing as low a note as you possibly can, especially if you can make it resonate in your throat. This also has the effect of sending vibrations through your head, but in this case the vibrations are periodic. This leads to some very interesting results, with time-varying light sources appearing to cycle slowly in space! Adjust the pitch of the note you're singing until you see this effect.

What you're looking for: the "retinal trace"

Whether you're jiggling your eyeballs or using a mirror or other technique, what you want to see is a trail of light across your retina.

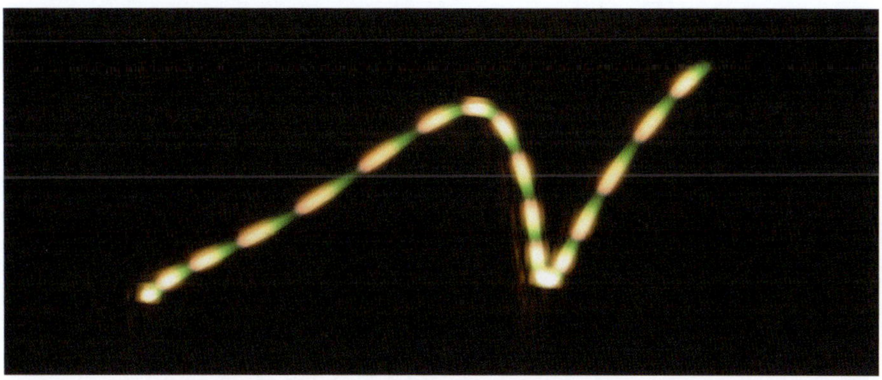

Figure 17: This photograph, which was created by moving the camera while the shutter was open, shows a trace such as you might see when vision jiggling. In this case, it's an outdoor light on a building.

Because the image of the light source is travelling across your retina, this trail can be regarded as a graph of the light's intensity (and/or color, and/or spatial pattern) against time. It's not as neat and tidy as a graph with x and y axes, but it does contain everything you can observe about the time-varying nature of the light source.

Because you're now an observer of a time-varying light source, the retina's persistence of vision is working *for* you rather than against you, because it allows the retinal trace to remain visible for a short time – long enough to see it and interpret it. With practice, you will be able to identify the features of a retinal trace and interpret them. Most of this Field Guide is devoted to helping you to do just that.

Tips for a good experience

Do it in the dark

It should be fairly obvious that you will get the best results by making your observations in the dark, either at night or in a dark location, such as a room with the lights turned off. This will increase the visibility of the trace on your retina.

Note that sometimes it's possible to make observations in daylight, as long as the light source you're interested in is bright and distinct enough. A great example is a water drop I observed in sunlight. (See page 147 for the whole story.)

Accounting for the shape or arrangement of the light source or sources

Figure 18: A well-known sign with a distinctive shape.

Some light sources, such as distant street lights, appear to the eye as simple dots, and leave clean, narrow traces. However, if the light source has an obvious shape, such as, say the "M" logo of a McDonald's restaurant, then the trace will be "painted" by this shape.

This is not usually a problem; and in fact it gets particularly interesting if the light source flashes with discreet pulses, such as an LED tail light cluster on a car. The circular form of the GMC Acadia's rear lights is especially rewarding to observe (Figure 20, and see page 59 for more details).

Figure 19: A photographic trace of a McDonald's sign. Note how the trace is a combination of two things: (a) the shape of the sign, and (b) the time-varying behavior of the sign, which accounts for the subtle periodic changes in brightness along the trace. (All the other traces in the photograph come from other lights in the scene.) (1/4 sec)

The shape of a light source, or the arrangement of a group of light sources can sometimes have a detrimental effect on your ability to observe the time-varying elements of the light. For example, an outdoor LED display that shows time and temperature may be hard to observe if it's surrounded by other bright signs, as these will also leave traces that may overlap the trace you are trying to observe. Masking (page 20) may help.

Finally, when interpreting complex traces it is sometimes useful to think of the light source as a "paint brush". This can help you to interpret whether what you see in the trace is a result of the light's time-varying behavior or its shape. The answer is: it's both, and you need to account for that.

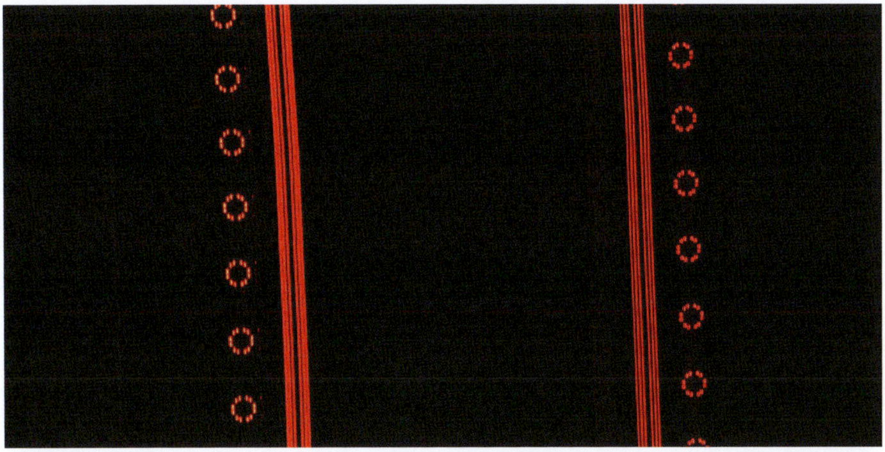

Figure 20: A trace of the tail lights of a GMC Acadia. Note how the circular form of the time-varying LED clusters gives rise to a trace of circles. Note too that these circles are not stretched or smeared in the direction of the trace. This indicates that the "on" time is very brief. (1/10 sec)

Direction of jiggle

A consequence of the above is that sometimes an up-down jiggle will not work as well as, say, a side-to-side jiggle. Imagine a building with a series of lights down the side. The lights would be in a vertical column, one above the other. If you were to jiggle your vision up and down, the trace from each light would lie on top of the traces of all the other lights, and you would see a confusing muddle. To fix this, simply jiggle side-to-side, and this will separate out the individual trails on your retina. (People who have not mastered side-to-side jiggling may need to tilt their heads to the side.)

Figure 21: A scene with a single fluorescent tube light.

Figure 22: Sweeping your vision *perpendicular* to the length of the tube produces a clear trace with the time-varying behavior clearly visible. (1/10 sec)

Now imagine a fluorescent tube light. If you were to jiggle along the axis of the tube, the trace would be obscured by the extent of the tube itself, leading to poor separation of the phases in your retinal trace. Instead, jiggle perpendicular to the tube and you'll see a broad trace with very good separation of the time-varying features.

Figure 23: However, sweeping *along* the length of the tube produces a single bright line with no detail. (The yellow dashed line is a street-light beyond the building.) (1/10 sec)

Speed of jiggle

How fast should you move your eyes (or mirror)? The usual answer is: as fast as possible. The reason is simple: The faster your eyes are moving, the faster the image will traverse your retinas. This results in the features of the trace being separated further from each other, and therefore easier to discern and interpret. Occasionally, however, a slower speed – or a smaller sweep – might be appropriate, especially in cases where the light source is dim, as this helps to prevent the trace from becoming too faint.

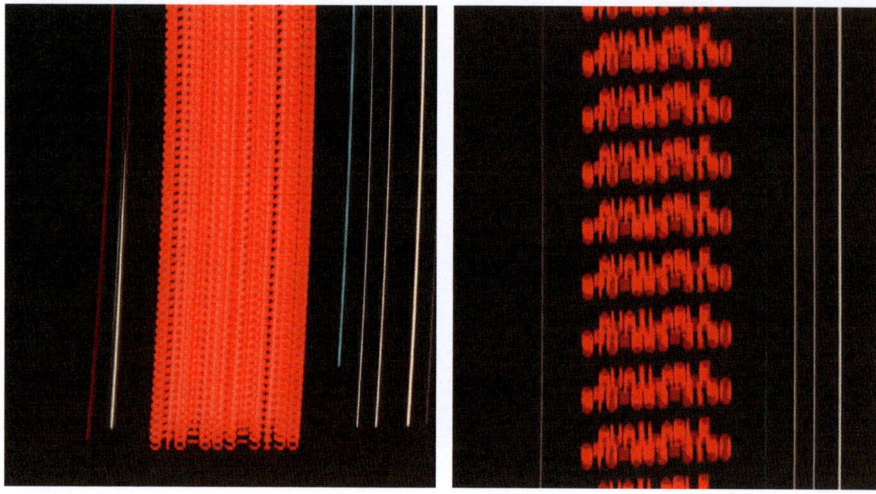

Figure 24: Two photographic traces of an LED matrix sign. This sign was flashing very quickly, maybe 1,000 times per second. When taking the left-hand image, the camera was not moved fast enough, so the individual images are not separated enough to see the detail. However, for the right-hand image the camera was swept much faster, allowing the individual images of the sign to separate. Now we can see more detail. (Both 1/10 sec)

Brightness

Some light sources are too bright to make a good retinal trace. The retina becomes over-stimulated and even with your fastest jiggling it's not easy to separate out the phases. Ways to diminish the brightness include wearing sunglasses, placing yourself further away from the light source, or placing yourself so that the light source is shining less brightly in your direction. The last approach applies well to office projectors (page 93).

In other cases, the lights may be too dim to generate discernible traces. This might be due to distance, such as street lights that are several miles away. If possible, move closer to the lights.

Masking other light sources

Sometimes the light source you are interested in may be in close proximity to other light sources, and if you try to observe them with eyeball jiggling or a mirror the other light sources may make it hard to see the retinal trace you are looking for. In these cases you may need to mask the other lights, for example with your hand. Forming a tube with your hand is often a good way to isolate the light source you're interested in, though it's hard or impossible to do when using a mirror or a camera.

Figure 25: Some suggestions for how you can use your hands to mask unwanted light sources out of your line of sight.

A technique that may work when using a mirror: hold the mirror further from your face. This reduces the field of view and can serve to mask out unwanted light sources around the source of interest. This is best done by finding the light source with the mirror close to your face and then slowly moving it away, taking care to keep the light source in view. Then you can start jiggling.

Distance

In general, you should avoid standing too close to the light source you're observing. Being too close can give rise to two problems:

- If the light source is large, such as a sign, then you may not be able to jiggle up and down far enough or fast enough to separate the phases of the light in the retinal trace. In fact, the further away you are, the more separated the phases will appear, because the light source itself will appear smaller on your retina. For the geometrically minded, we're talking about *angular* size, i.e. the angle that the light source subtends in your view. Reducing its angular size makes it much easier to separate out the features in the trace. It's like painting with a finer brush.
- If you're too close, it's possible that the light source is too bright, and will saturate your retina, and you will lose detail in the retinal trace, as discussed above.

Moving further away will alleviate both of these problems.

Conversely, standing too far from a light source may make the trace too small or faint to be interpretable.

Single sweep jiggling

CRT televisions, computer monitors (see page 81) and some other sources exhibit complex time-varying behavior, and the direction in which you sweep your gaze matters considerably. A single downward sweep, for example, may look completely different to an upward sweep (both are interesting). Left and right sweeps can differ too. In any case, repeated single sweeps (instead of regular back and forth jiggling) can help you to interpret complex information.

Observing complex light sources

In some cases, the light source is so complex that if you jiggle back and forth you can be overloaded with information and won't be able to interpret everything that's happening. Some dot matrix signs (page 106) are like this.

Don't worry – try a combination of single sweeps and back-and-forth jiggling and take your time to understand what you're seeing. Look, think and look again, and it may eventually make sense. Alternatively, take some photographs (page 27), which will allow you to study the details over a cup of coffee. In fact, photographs often provide details that you don't catch when vision jiggling.

False positives

When observing a particularly bright light source, such as a car's headlight, you might possibly imagine that you can see small but sharp interruptions in the trace, and therefore come to the conclusion that the light source is time-varying. However, this might be a case of being too dazzled to see clearly. If you happen to look directly at the light for a second *before* jiggling your vision, the point on your retina where the light was resting will become desensitized. (This point is the fovea, which is the most sensitive and detail-capable part of your retina, and is what you "look at things" with.)

Usually, this desensitized spot would appear to you as an after-image that you can see for a while (imagine what you see after glancing at the sun). When vision jiggling, however, this desensitized region shows up as a dark spot in the trace, every time the trace passes through your fovea. If you suspect this might be happening, look away for a minute and then jiggle again – *without* allowing the light source to sit still on your fovea for any length of time.

Finally, I have occasionally seen a "trail of dashes" out of the corner of my eye only to find on inspection that it was actually a static row of identical lights.

Overcoming the brain's suppression

If this chapter has helped you to see retinal traces, you might be wondering why you had not seen them before. After all, the optical system of the eye is very capable of seeing them, so how come one doesn't see them all the time?

I believe the answer comes down to the way the brain works – or, more properly, the eye *and* the brain. The retina is in many ways an extension of the brain and applies significant information processing before sending the vision signal to the visual cortex. The retina detects edges, areas of contrast, areas of motion and colors, among other things. Then the visual cortex correlates this interpreted vision with knowledge held elsewhere in the brain before presenting it to your conscious perception. This process helps you to very quickly see *things* and *spatial relationships*, rather than just patches and shapes of color. In other words, in many ways you see the world according to how you already understand the world.

Retinal traces are a figment of the eye resulting from the retina's persistence of vision. They are not part of the world in front of you. This is something that your brain learned early in your life, and so it has a tendency to label them "unimportant", i.e. to be ignored. Therefore, when you see the world at night, with cars and street lights, although your eyes are often generating retinal traces, you don't *see* them because your brain suppresses them from the set of "things to pay attention to".

I spoke to a lady whose brain works a little differently. She was very excited when I told her about my book, because she has seen retinal traces for years, and has regarded them as a nuisance because her brain wasn't demoting their importance. She thought she was the only one who saw them. When I told her I was writing a book about them, she learned that rather than being a nuisance, retinal traces are something that can be enjoyed. She now enthusiastically tells me about what she has seen.

After all, they provide true information about the world that you would not otherwise see, and there's nothing wrong with that.

Part of the process of learning how to observe time-varying light sources is to train your brain that the traces *are* something that merits attention. It's the same kind of training that an audio producer goes through to learn how to hear and adjust EQ in a music mix, or when a photographer visually scans a scene to identify challenges with dynamic range. These are *learned* forms of specialized perception.

Once somebody helps you to see traces, you will start seeing them everywhere.

We will see on page 27 that the behavior of the eye and brain together has additional consequences.

Aren't traces a distraction?

I can only talk about my personal experience when I say no, retinal traces are not a distraction. I have learned to pay attention to them when I want to and to ignore them when I want to. In fact, the pleasure of seeing an interesting trace serendipitously is one of the highest joys of eyeball-jiggling. Go out there and enjoy those traces!

Retinal traces and how to interpret them

Intensity over time

Almost all time-varying light sources vary in intensity (brightness). In many, but not all, cases, this is simply a consequence of the AC (alternating current) electrical power supply, which repeats its cycle of voltage 60 times a second in North America or 50 times a second in Europe. Sometimes both halves of the AC cycle are used so the light source's intensity will peak 120 (or 100) times a second. Don't worry: 120 times a second is easy to observe with the techniques described in this book. You will be able to observe light sources which flash thousands of times per second!

The following image shows a photographic trace for a neon indicator, in this case on an electrical power strip.

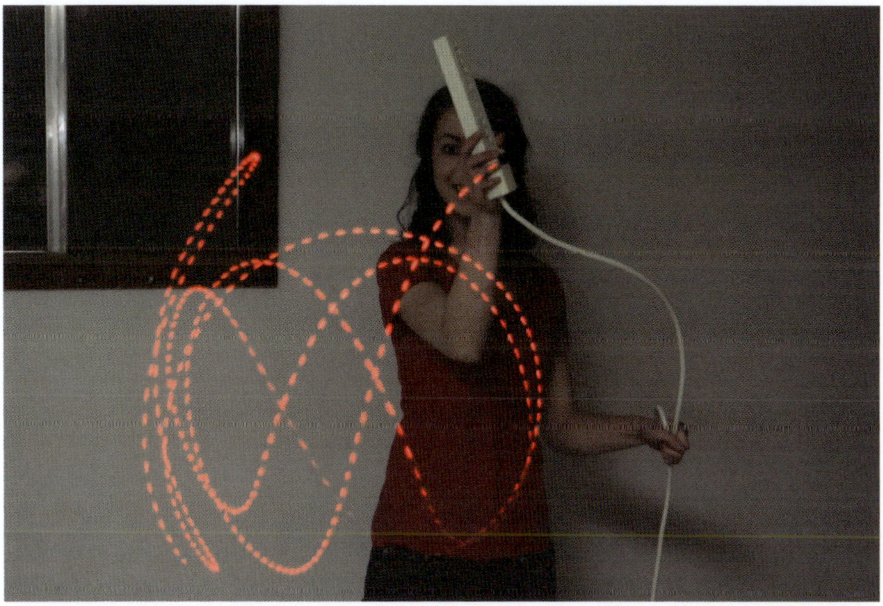

Figure 26: A trace of a neon indicator. This is a photograph produced by "light painting" - see page 41. (2.5 sec)

You can see a trail of orange dashes separated by dark gaps. In this photograph, the red dashes are about the same length as the gaps. Already we're able to observe that instead of shining with a steady light, this indicator is actually only shining for 50% of the time.

Notice also how the ends of the individual dashes are a little soft. This shows that the brightness rose and fell somewhat slowly, rather than abruptly turning on and then off. If this were an LED controlled by a digital circuit, the ends of the dashes would probably be very abrupt (see Figure 176 for an example of this). Here we can see that neon lamps driven directly by the AC voltage don't behave like digitally-controlled LEDs.

We're already interpreting a trace and learning something from it!

Figure 27 once again shows the trace of the tail lights of a GMC Acadia. Ignore the unbroken traces; these were generated by the inner tail light, which is DC powered and therefore

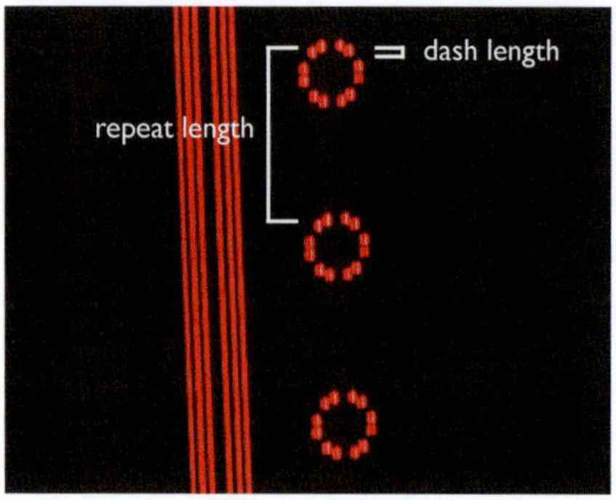

Figure 27: A photographic trace of the tail lights of a GMC Acadia. Here, we have zoomed in to consider the detail.

not time-varying.

There are two things to notice: First, instead of the neat dashes seen in the previous figure (which arose because the neon indicator is more or less dot-shaped), we now see the circular shape of the tail light itself (we discussed this on page 18).

Secondly, if we assess the length of the repeat relative to the length of the dash created by any given light emitting diode (LED), we can conclude that the tail light is shining with red light for less than 10% of the time! Seen on the street, it may look like a steady red light, but that's an illusion. We'll talk more about car tail lights on page 59.

Notice that the transitions between the light and dark states are pretty much instantaneous – there's no fade-in or fade-out. This is typical of digitally-controlled LED lighting.

Finally, here's a photograph emulating a retinal trace for a pressurized sodium street light, which is the most common type of street light in many countries (see page 50 for more information).

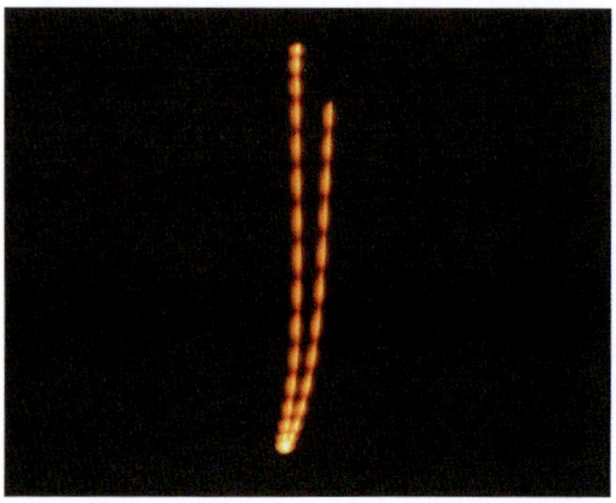

Figure 28: A photographic trace of a pressurized sodium street light, taken with an up-then-down sweep of the camera. (1/4 sec)

Note that this light also passes through bright and darker phases, but that the transitions between bright and dark are much more gradual. Your eye would usually also see an apparent thickening of the retinal trace in the brighter areas, similar to this image.

These three examples illustrate some of the variations in intensity that can be observed.

Color over time

Another characteristic of light sources that can change over time is their color.

I've written more than once about fluorescent tube lights in the early sections of this book. Now it's time to pay another visit. In Figure 29 you can see a broad band of light. The width of this band is simply due to the length of the fluorescent tube. The most interesting feature of this graph is the change in color – slightly bluish white alternates with a dull yellow, which has at least two levels of brightness. Again, it's easy to make a judgment of how much time is spent in each phase: in this case, it's around 50% in the white phase and 50% in the yellow phase. Fluorescent lighting is covered in detail on page 70.

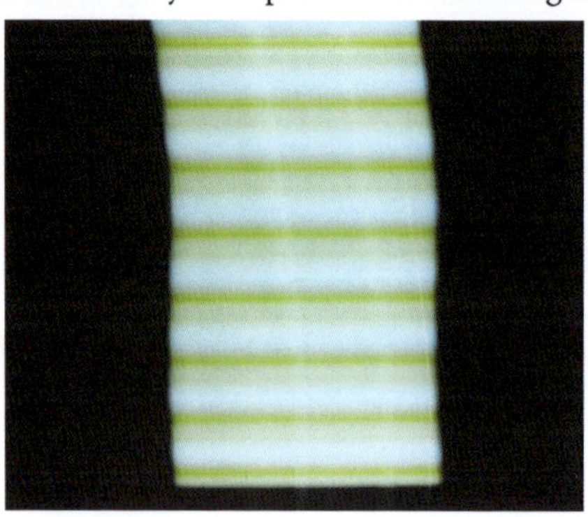

Figure 29: A photographic trace of a fluorescent tube light. (1/10 sec)

Most simple light sources don't exhibit such a strong variation in color. However, I did once see a light hanging above a table in a Denny's restaurant in Livonia, Michigan, which exhibited an incredible emerald green phase that was totally invisible without eyeball jiggling. While writing this book, I was lucky enough to find another such light at a motel in the Black Hills of South Dakota – see the following figures.

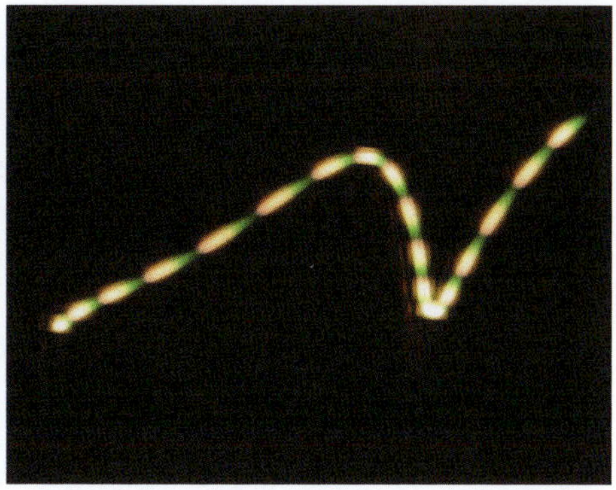

Figure 30: *(left)* An outdoor lamp at a motel in South Dakota.

Figure 31: *(right)* A photographic trace of the lamp in the previous figure. Unlike all the other lamps along the front of the motel, this one displayed the wonderful emerald green phase that you can see here. Without using the techniques in this book, the emerald phase would be completely invisible as it's overwhelmed by the brighter apricot-colored phase. (0.3 sec)

Patterns and what they tell you

Although a single light source can usually show only intensity and/or color variations, many displays consist of multiple individual light sources. Think of the blue or green numeric display on a DVD player or microwave oven. There's a separate light for each segment of each number, and others for the various little indicators. What's interesting (to the observer of time-varying light sources) is that it's usual for the numbers or individual segments to be illuminated at different times.

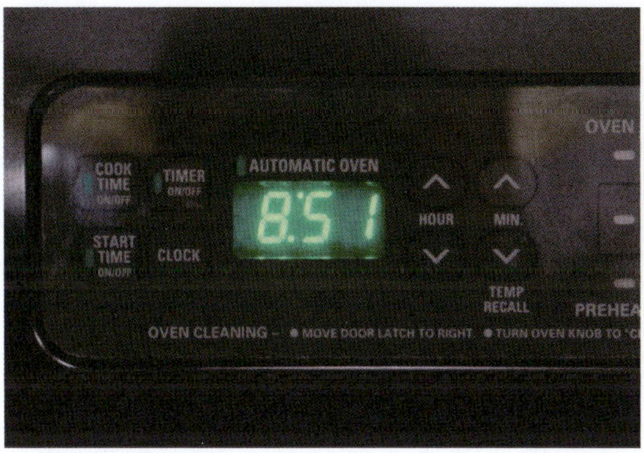

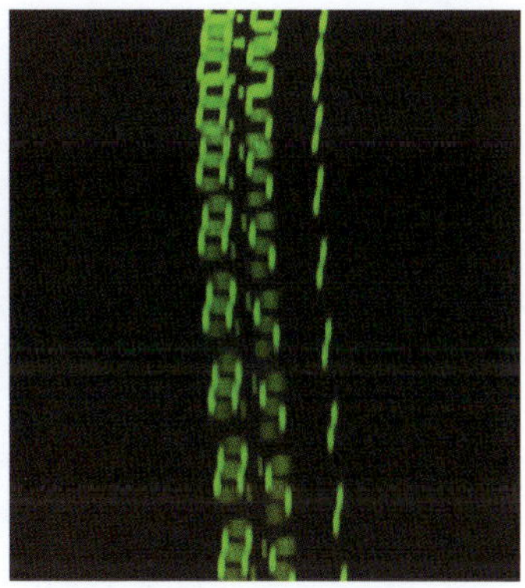

Figure 32: *(left)* A typical numeric display on an oven.

Figure 33: *(right)* A photographic trace of the display, captured by sweeping the camera downwards when shooting. You can see how the three numbers are no longer lined up next to each other. This shows that they're not illuminated at exactly the same time. (1/10 sec)

Numeric displays are covered on page 79.

This was a simple example; more complex examples are given in the Field Guide. As you'd expect, the more complex the pattern, the more work it will take to interpret the light source's behavior.

Comparing light sources

Sometimes it's interesting to observe multiple light sources at the same time to see if there are any differences in their behavior. Probably the easiest case to observe is two or more street lights to determine if they're on the same phase of the "three-phase" electrical supply. This observation is covered on page 51. Another observation is to see if one light out of several is behaving anomalously, which could indicate that it's going to fail. The light in that Denny's restaurant was a good example.

When comparing the *timing* of multiple light sources, it's sometimes important to choose the correct jiggle direction, to avoid falsely interpreting how the traces line up. This is explained in detail in the section on pressurized sodium lighting, on page 52.

Figure 34: *(left)* Here is an example where we can compare the timing of two light sources. This is a charger for rechargeable batteries, with a red indicator LED above each battery. To the casual observer, it appears that both LEDs are simply "on". (1/160 sec)

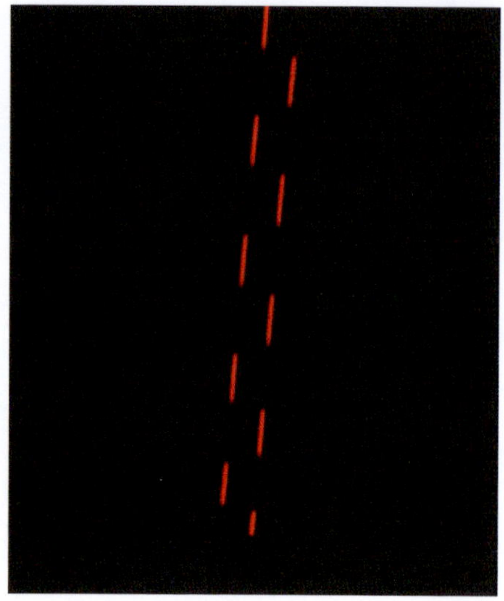

Figure 35: *(right)* A photographic trace of the battery charger, captured in darkness. The alternating dashes (which are easy to see with vision jiggling) reveal the true behavior of the two LEDs. Not only are they *not* constantly illuminated (they are on for a little less than 50% and off for a little more than 50% of the time), but they are in "antiphase", meaning that when one is on, the other is off and *vice versa*. Why? To find out, we would have to ask the engineer who designed the electronic circuit. (1/4 sec)

Figure 36: To demonstrate that only one LED is illuminated at any time, this photograph was taken with a short exposure time. During the exposure, the right-hand LED was off. Both LEDs appear to be illuminated in Figure 34 because the exposure time was longer and both LEDs spent some time illuminated during the time that the shutter was open. (1/250 sec)

Photography

Photographing time-varying light sources can be as rewarding as observing them with your eyes. By moving the camera during the exposure, the image of the light source can be made to travel over the film or (for digital cameras) the light sensor, and the results are similar to what you would see by eyeball jiggling or using a mirror. The advantage of using a camera, of course, is that the image is permanent and can be used in, for example, this book. It also allows deeper analysis, such as measuring exactly the length of each phase as a percentage of the complete cycle, or seeing detail that can't easily be discerned from a transient retinal trace. Finally, it's possible to capture visually striking art images.

Eyes and cameras are not the same

Be aware, though, that a camera does not behave in exactly the same way as the human eye. There are two main differences, and I will highlight them again at the beginning of the Field Guide on page 47. They are:

- The color of a bright light source can be washed out to white (or even a different, completely wrong color) much sooner on a camera than in the human eye.
- The human eye does not blur everything in the way a camera does when jiggling; it tends to preserve most of the scene apparently unblurred.

The issue with color is covered in more detail later in this section (specifically on page 33); but I would like to go deeper into the second item here.

Naturally, when you jiggle a camera to capture a trace of a time-varying light source, anything that is in the scene will be blurred. After all, that's what happens when you don't hold the camera still when taking a photograph – it's called "motion blur" or "camera shake". But human vision is different. Believe it or not, when you're eyeball-jiggling (and to a lesser extent, when mirror-jiggling), you will see the scene mostly intact. An example is a street light in front of a sunset: you will see the sunset just fine, and the trace will *appear* to overlay it. This is because the eye and brain are much more sophisticated than a camera, and can integrate elements in a scene into a coherent whole.

I like to teach people that the mission statement of the eye and brain is "to make sense of the world", which means that what you actually *perceive* has already undergone substantial interpretation in your retina and visual cortex *before* it reaches your conscious experience. (This, of course, is why optical illusions happen, where what you "see" can be very different from what is really in front of you.)

A camera cannot do this; it simply captures how the light fell on the sensor while the shutter was open. (Some cameras do apply some level of "intelligence" when generating JPEG photographs, such as in-camera red-eye correction, but this doesn't apply to photographing time-varying light sources.)

The following images attempt to illustrate the difference. All are captured with a camera (of course); but the image that illustrates what you might see when vision-jiggling is simulated using Adobe Photoshop.

Figure 37: A photograph of a boy wearing a cycling helmet with a white LED headlamp.

Figure 38: A simple swept photograph of the same scene. Notice that everything is equally blurred vertically. Only the traces from the lights are clear. (1/10 sec)

Figure 39: This image is composed from the top photograph, an additional photograph of the boy, taken lower down in the field of view, and a swept photograph of the headlamp. I used Adobe Photoshop to compose an image that's *somewhat* representative of what you actually *perceive* when vision jiggling. Notice how the traces are clear, *and* how it's possible to see the scene. It seems that the brain takes advantage of the moments when the eyes are static at the top and bottom of the oscillation, and holds on to the perception of the whole scene.

What type of camera do I need?

This section relates to digital cameras. I would expect film cameras to behave in much the same way, but I haven't had the opportunity to experiment, or in fact much interest in doing so. Digital cameras have one great advantage: you can take many pictures without thinking of the cost of the film! Indeed, when trying to photograph time-varying light sources, it's often necessary to take a lot of pictures in order to be sure of having some really good ones. In preparing this book, I took at least 3,500 photographs!

Another advantage of digital cameras is that the LCD display on the back lets you review the picture immediately. This is extremely valuable for this type of photography because you can take a picture and judge if any adjustments are needed to your sweeping technique or the settings of the camera. It allows you to keep shooting until you get the images you're looking for. With a film camera, you wouldn't know until you develop the film how well the shoot went.

A larger display on the back of a digital camera is better than a small one. When I moved up from a Canon EOS 20D, which has a 1.8" display, to an EOS 40D (3.0" display), I certainly appreciated the improved ability to judge the success of each shot.

On the down side, digital cameras tend to have a lower dynamic range than film cameras, which means that the ability of the digital camera to capture the whole range from dark to light is not as good as for a film camera. I don't think that this matters here; but die-hard film photographers are welcome to experiment. Another thing that film users would need to consider is the ISO rating of the film used. I would suggest a relatively low ISO, such as 200 to 400. This may seem counter-intuitive given that one is photographing dark scenes; but the objective is for the *lights* in the scene to be captured, and for the scene itself to disappear by virtue of being under-exposed. We'll discuss this in more detail below.

I have not had the opportunity to experiment with the new class of shutterless digital cameras. Some of these have what is called a "rolling shutter", which is where each row of pixels is stored to the image in sequence. This means there is a period of time during which the image is captured from top to bottom. For a normal scene or a portrait, this doesn't present a problem. But when capturing time-varying light sources, it may possibly introduce time-related artifacts which can render the traces hard to interpret. Having said that, cellphones often have this type of technology, and I have successfully captured a few traces, though with mediocre quality since most cellphone cameras cannot be set to a manual exposure time, and can get confused about focusing. Your mileage may vary.

Regardless of whether film-based or digital, the camera you use should allow full control over exposure. Exposure consists of three parameters:

- shutter speed (or exposure time): this determines the length of time during which the shutter is open, allowing light to reach the sensor or film;
- aperture: this determines the size of the iris which controls the amount of light that reaches the film or light sensor;
- ISO (sensor sensitivity): this contributes to exposure: more on this later in this section.

For best results, the camera should allow full, independent control of all three parameters. Usually, this is indicated with a mode called **M**, for "manual". Modes such as **P** (program), **A** or **Av** (aperture priority) and **S** or **Tv** (shutter-speed or time priority) are less likely to give good results.

The technique I have used most often is to set the camera to **M** mode, and the exposure time to 0.1 seconds (1/10 sec). Then I can use the aperture setting to adjust the exposure, usually to significantly under-expose the photograph.

Cameras which do not allow the user to control the shutter speed are harder to use for this kind of photography. This means that most cellphones would serve only poorly. However, with care, quite good pictures can be obtained. Figure 244 on page 116 was taken with an LG G2 cellphone, and Figure 320 on page 152 was taken with an HTC Rezound.

Figure 40: The Canon EOS 70D digital SLR camera. (Photograph: Martin Kraft)

Finally, if your camera has image stabilization technology in the lens or the body, I recommend switching it off. Image stabilization only works for small angles of camera movement, and won't do any good for the large angles covered in a swept or jiggled photograph; and anyway, you're not looking for a stable image!

Most of the photographs I took for this book were taken with Canon EOS 20D, EOS 40D, EOS 60D and EOS 70D digital SLR cameras.

How to photograph time-varying light sources

When photographing time-varying light sources, your objective is to "paint" a trace across the camera's light sensor, using the light source as a "brush". In some cases, the light source may be moving (example: Figure 26), but in most cases the light source is stationary so you will have to move the camera.

Clearly, if we're trying to capture the behavior of a time-varying light source, we need to hold the shutter open long enough for the camera to "see" everything that the light source does during at least one cycle. For a 60Hz light, this would mean a minimum of $1/60^{th}$ of a second. Figure 41 is an example of an exposure that is too short.

However, it's much more interesting to hold the shutter open for longer, so that a trace is produced (Figure 42).

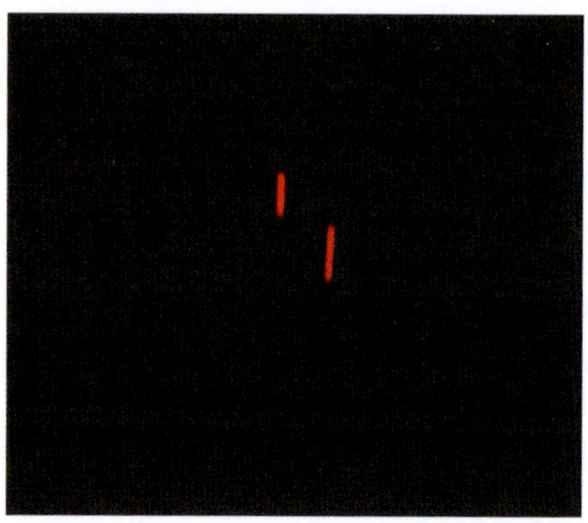

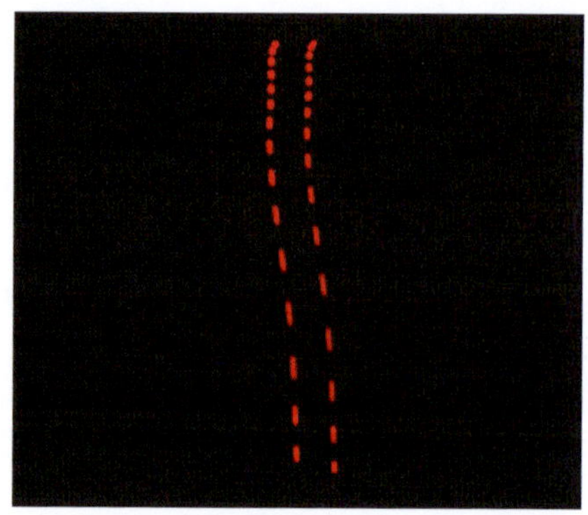

Figure 41: *(left)* Two adjacent LEDs with a 60Hz cycle, captured with a 1/80 second exposure. It's not possible to see any periodicity, and so we can't be certain of the behavior of these light sources. (See Figure 34 to see the light source shown here.)

Figure 42: *(right)* The same LEDs captured with a 1/4 second exposure. While maybe a little too long for a purely documentary image, the result is attractive and informative.

Here are some steps which will help you to maximize your success with time-varying light source photography:

- Experiment!
- Select a low to moderate ISO, such as 400. Remember, you are trying to capture the *lights* in the scene, not the scene itself, so you should aim to under-expose aggressively. (This also helps to avoid colored lights washing out towards white in the photograph; see page 33.)

- Select a shutter speed of 1/10 (one tenth) to 1/4 (a quarter) second. This will result in a clear, uncluttered image in which the traces of the lights will be clearly seen. Most of the photographic traces in this book were taken at 1/10 second. If you're making artistic images, feel free to expose for longer, to create more complex traces. Experience will help you to determine a suitable exposure time.

- Prepare by setting the focus. If your camera uses auto-focus, then you need to be focused *before* you take the photograph, because the auto-focus system will not be able to lock when you're sweeping the camera across the scene. Most cameras allow you to focus on a scene by pressing the button half-way. Once it's in focus, don't release the button, but take the photo as described in the next step. Alternatively, switch to manual focus.

- Before taking the photograph, develop a sweeping motion that will deliver the shape of trace you're aiming for. For example, a quick up-then-down sweep with a slight sideways drift will give you a trace like the one in Figure 43.

  This is useful for clearly showing the behavior of the light source while at the same time emulating what eyeball-jiggling would look like.

- While sweeping, take the photograph, trying not to pause in your sweeping motion. For documentary photographs especially, a smooth motion will result in an easily-interpretable trace.

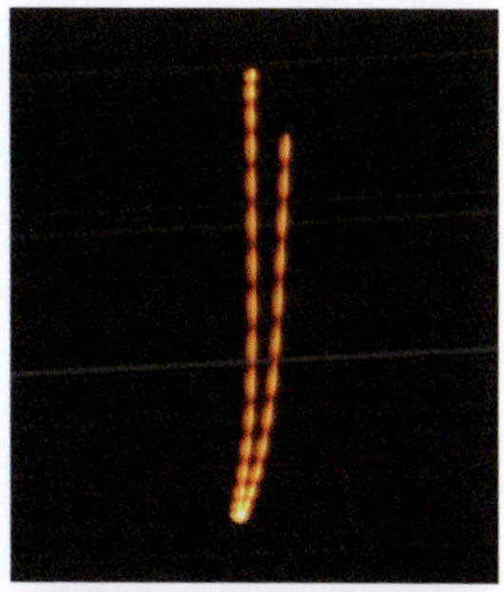

Figure 43: A photographic trace of a single pressurized sodium street-light. (1/4 sec)

- Some light sources, such as a vertical neon tube, will require a sideways sweeping motion to capture the trace (see Figure 44).

- If you are creating art, use your sweeping motion like a paintbrush. Experiment!

Figure 44: A vertical light requires a horizontal sweep. This is an accent strip on the corner of a building, illuminated by white LEDs. (1/10 sec)

- If you are using a digital camera, review the photograph on the display on the back of the camera. If the image is too dark, open the aperture a little (select a lower "f" number). If it is too bright, and the lights are washed out or you don't see good colors, then close the aperture a little (select a higher "f" number) and try again.

- You may also find that you need to adjust the speed of your sweeping motion, or the direction, or the number of sweeps, or the smoothness. You will probably find that you sometimes miss the light source completely! Consider whether you want the trace to be completely contained in the picture, or to go out of the frame (there are reasons for both; we will discuss this shortly).

- Take many pictures of each scene so that you'll find the one or two that are exceptionally good.

- Don't be scared to optimize the photos "in post" (professional short-hand for "in post-production") using applications like Photoshop, Lightroom or The GIMP (a free image editing application from www.gimp.org). Enhancing the exposure can make an image clearer and more arresting. Many of the photos in this book have had some such adjustments made.

- Experiment! (I seem to recall saying that before...)

Image information: EXIF

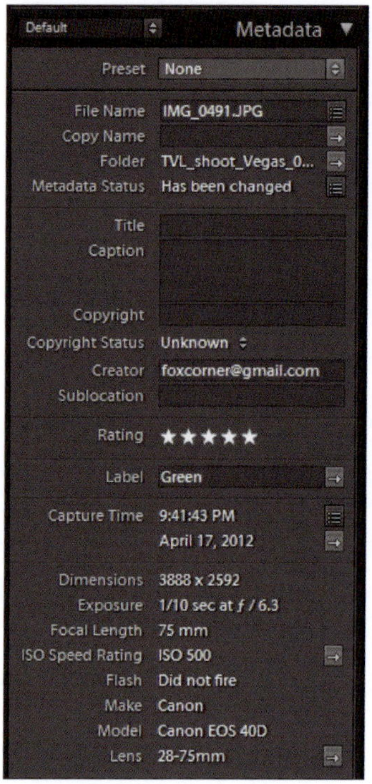

Figure 45: This panel shows the most important EXIF data for a particular photograph, as presented in Adobe Lightroom. Note especially the "Exposure" information, which in this case shows that the photograph was taken with a tenth of a second exposure time.

When you load the images onto a computer, you should be able to use your photography software to inspect the EXIF (Exchangeable Image File Format) data, which was recorded by the camera and incorporated in the JPEG image file. RAW images also carry the same data; software that knows how to handle RAW files will reportt this information. This will tell you what the camera settings were when the photograph was taken, including aperture, shutter speed and ISO. You will be able to see what worked best during your photo shoot.

JPEG vs. RAW

Most professional photographers "shoot RAW", which means that the camera is set up so that it saves a RAW image file for each photograph taken. A RAW file is essentially a capture of the contents of the camera's image sensor, and therefore offers the best scope for getting the best image quality (assuming you are using software that can process RAW files).

Not all cameras offer RAW capability. However, all digital SLRs do.

JPEG images, on the other hand, have already undergone extensive processing by the camera before being saved to the memory card. This results in smaller files which are viewable in any image viewer software. You can fit many more JPEG photographs onto a camera's memory card than you can RAW images, so if your card is low on space, select JPEG.

All cameras can save images as JPEG files.

For time-varying light source photography, there is no strong reason for selecting RAW or JPEG. Many of the photographs that I took for this book were captured as JPEG files. However, RAW files can provide greater scope for optimizing the exposure or contrast of the images in your editing software, which may prove useful in particularly difficult lighting conditions. If you are aiming to create high quality art prints, shooting RAW is recommended.

Measuring the rate of variation

It is possible to measure the frequency (flash rate) of a time-varying light source by taking a photograph, inspecting the EXIF to determine the exposure time, and counting the bright pulses along the trace. For example, if your shutter speed was a tenth of a second, and there are fifteen bright dashes, then the light was flashing 150 times per second (150 Hz). With care, it may be possible to measure the "on" time relative to the "off" time; however, don't forget to account for the shape of the light source (page 18).

In order to make frequency measurements, it is important that the start and end of the trace are both within the picture. If this is not the case, there is no way to know how many flashes occurred during your known exposure time, since some of them were out of the picture and therefore not recorded. The following figures illustrate this.

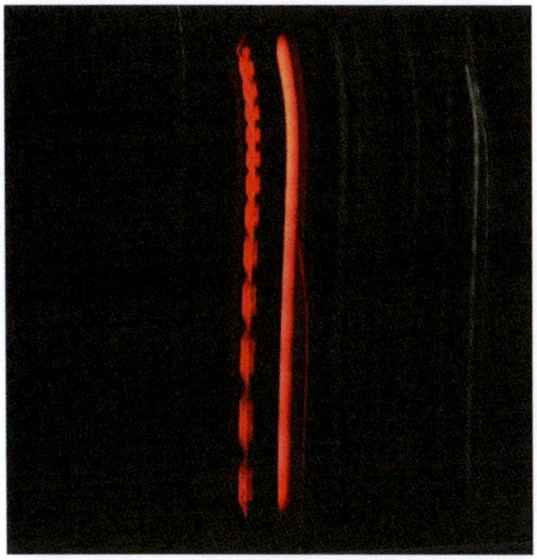

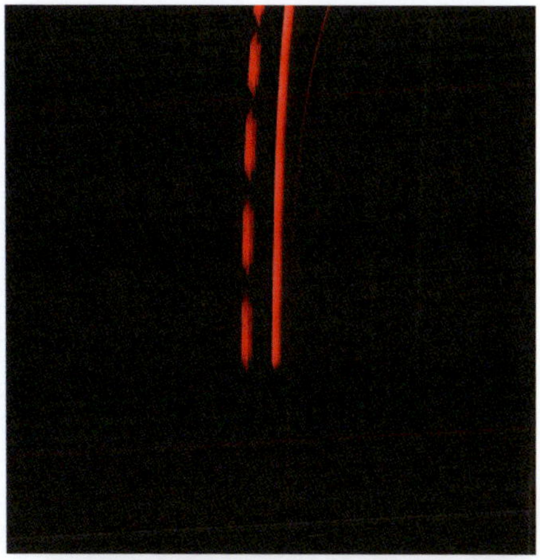

Figure 46: *(left)* This is a photographic trace of a pair of stop lights, of which the left light (an arrow) is strobing (see page 65). It was taken with an exposure time of 1/10 second. Close inspection of this image shows twelve flashes of the light, which gives a flash rate of 120 flashes per second, or 120 Hz.

Figure 47: *(right)* This picture is not usable for determining the flash rate because the trace goes out of the picture, so it's not possible to count how many times the light flashed during the 1/10 second exposure.

Exposure and color

Always bear in mind that even the best digital cameras do not behave exactly like the human eye. In particular, the color of a bright light source can be shifted substantially by the camera. For example, red can move towards yellow on Canon cameras, whereas on Nikon cameras red tends to move towards white. To minimize this effect, try under-exposing aggressively (lower ISO and/or smaller aperture).

You can often recover the brightness of the image in a software application such as Adobe Photoshop, Adobe Lightroom or The GIMP. However, this kind of editing can emphasize the camera's sensor noise, which is undesirable. Shooting with a lower ISO setting will reduce this noise. Newer cameras tend to have less noise in low-light photographs than older cameras do.

Note that a camera will be much less likely to over-expose lights in this way when swept. The "dose" of light reaching any point on the sensor will be less because the image is always moving to a fresh part of the sensor.

Just like anything else with photography, experiment to gain experience. In the end, the images you obtain may be so stylized or abstract that it may not actually matter if the color is not completely accurate. In fact, the subject of color accuracy from camera to screen to print is very complex and is outside the scope of this book.

The primary lesson here is that you should plan to underexpose.

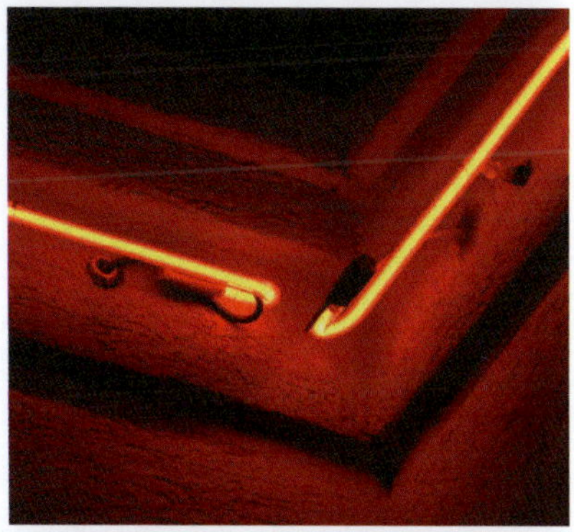

Figure 48: A photograph of two neon tubes. Although the photograph is pleasingly exposed, the bright red light has become yellow due to the way that the camera's sensor responded to the brightness. This yellow is not at all the same color as the original neon tube. The illumination on the walls, being dimmer, shows the correct red color.

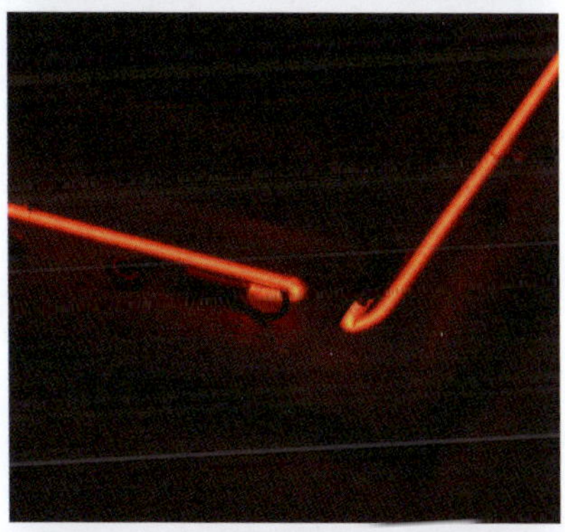

Figure 49: By under-exposing aggressively, the red color of the neon tubes is captured accurately. However, the surrounding scene is now much darker. For capturing photographic traces, this is desirable, because you want the light to make the trace, with no distraction from the background scene.

Figure 50: This is the same under-exposed photograph with some gentle brightening applied to the darker parts of the image using Adobe Lightroom. You may want to do this to optimize a conventional reference photograph that you would present alongside the photographic trace.

Taking a reference image

If you are capturing images for documentary reasons, it's a very good idea to take a conventional picture of the scene so that you can see each light in context. I call this a "reference image", and it will help you to understand the traces in the swept images. For example, study the following two photographs:

Figure 51: A night-time street scene in Greeley, Colorado. This photograph serves as a reference image for the next picture. Notice the apparent reflections of the brightest lights in the road. These are actually reflections inside the camera's lens.

Figure 52: A photographic trace of the scene in the preceding picture. Here are some features to note: The stop lights were red in the first picture, but are green in this trace. Look carefully for the three nearer green lights, which are clearly separate, and the group of green lights from further down the street, which together make a green brush-stroke. Especially interesting is the row of street-lights in the middle of the scene. In the trace, they are clearly shown to be time-varying, with dark red periods interrupting the predominantly orange illumination. These are pressurized sodium discharge lamps. What's interesting is that the dark bands of the entire row of lights line up neatly. This shows that all the lights are connected to the same phase of the three-phase electrical supply. See if you can figure out the correspondence between all the traces and the light sources in the scene. Hint: There's a short, bright trace at the top of the image which doesn't appear to have a source in the scene; but note that there's a lamp-post in the foreground, so presumably there is a street-light that is out of shot in the first image. (1/10 sec)

Zooming

Most cameras offer some degree of optical zoom (avoid using digital zoom – you might lose detail). Single Lens Reflex (SLR) cameras (like my Canon cameras) also allow the user to use different lenses, which offer different zoom ranges. So the question may arise: what is the best amount of zoom (properly: what is the optimum focal length) when creating photographic traces of time-varying light sources?

In general, the answer will vary for each photograph, and will be found by (you guessed it) experimenting. However, I would advise against zooming in too "close" to the light source for the following reason: the more you zoom in, the smaller the angle that you can sweep the camera before the light source goes out of the picture. It is better to zoom out somewhat and make a good sweep than to zoom in and be forced to use a constrained sweep. The reason for this is that you're aiming to separate out the changes in intensity, color and pattern of the light to give a clearly interpretable trace. This works best if you can sweep the camera through a fairly large angle when capturing the image.

Other considerations may be, for example, that you want to zoom out to capture a wide scene (all the streetlights on a street, for example), or that you want to zoom in to isolate one particular light source in a scene that's crowded with many lights. Or you might be creating art, in which case you should zoom according to your heart! In fact, zooming *while* taking the picture can give interesting results.

Figure 53: This image was created by zooming the lens during the 1/4 second exposure. The subject was a group of trees decorated with LED Christmas lights.

Side-effects of time-varying light sources on photography

In some circumstances, the time-varying nature of artificial lights can present a problem to the photographer.

I remember talking one Christmas Day to a friend of mine who is a professional photographer. He showed me a set of photographs he had taken with Nikon's high-end D3 camera. The photographs were of a gymnastics meet, and he was very pleased with the speed and sensitivity of the camera, which allowed him to take action photos indoors without using a flash.

I noticed, however, that some of the photos, say 20% of them, had a pink "cast", which means that the overall color was much pinker than it should have been. What was even stranger was that while some photos were fine, the pink ones came in three variations:

- the left part of the background was pink, but the rest of the background and the gymnast were fine;
- the right part of the background was pink, but the rest of the background and the gymnast were fine;
- and the gymnast was pink, but the whole background was fine.

Examples of those images are provided in the following figure.

We puzzled over this for a while, until I realized what was happening: The large discharge lights in the gymnasium were of a kind that generates time-varying light, with a dim pink phase between the bright white phases (see Figure 92). The camera was using short enough exposures (made possible by the D3's high sensitivity) that it was sometimes catching the illumination entirely during the pink phase.

Figure 54: *Top left:* An image with the expected coloration: no pink cast.

Top right: A pink cast on the left side of the image.

Lower left: Pink cast on the right.

Lower right: The background is mostly OK; the gymnast has a pink cast.

Photographs: Daniel Treat. used by permission.

Why wasn't it uniformly pink across the whole photograph? Answer: the lights were connected to different phases of the three-phase electrical supply, and were therefore producing pink light at different times in various parts of the building.

We then set out to prove this theory by going to his garage and taking several photographs under fluorescent lighting. This type of light has a pronounced yellow phase (easily seen by eyeball-jiggling or mirror-jiggling, see page 70). Sure enough, while most of the photographs were normal, several were strongly yellow-colored. Figure 126 and Figure 127 show this effect very clearly.

The moral of this story is that if you are taking photographs under artificial lighting that is strongly time-varying, you may get unexpected results. Take lots of shots to make sure you get the images you are looking for.

Another side-effect comes when photographing a scene which contains time-varying light sources. If a light source is flashing on and off very rapidly, it is possible to take a photograph with a short enough exposure time that the light is in the "off" state when the image is made. This might be a surprise if you were expecting the light to be "on" in the photograph. Look at Figure 36 for an example of this. The solution is to use a longer exposure time.

Finally, photographing CRT displays can give unexpected results. See page 81 for more details.

Post production

Another benefit of digital photography is that the photographs can be edited and optimized in "post production", which is a fancy term for what you do to the images on a computer. Even with the best shooting conditions and good parameters on the camera, there may be scope for enhancing the image to make it more informative or more visually attractive.

One of the most common operations that's worth doing is to increase the overall brightness and darken the blacks in an image. This causes the traces to "pop" better, and any noise or graininess in the dark background can be eliminated this way. I have applied this kind of processing to many of the images in this book. Note, however, that it's possible to take it too far: increasing the brightness can obliterate detail or cause colors to fade to white.

Adobe Photoshop's "Curves" tool is a great way to achieve these optimizations. Increasing the brightness can often be done very effectively by dragging the "white" end-stop to the left; and cleaning up the black parts of the image can be done by dragging the "black" end-stop to the right a little. In some cases, careful manipulation of the curve is what's needed. A good eye and some experimentation will help you achieve the best results for each image.

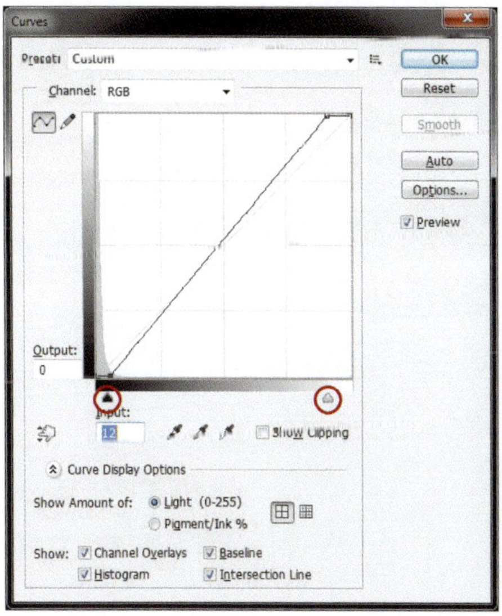

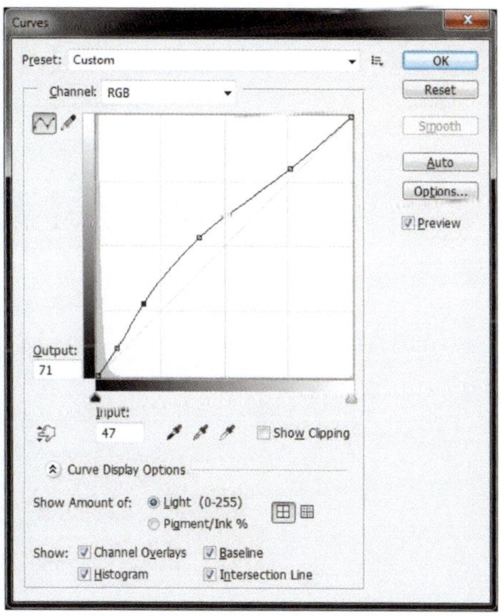

Figure 55: *(left)* The Curves tool in Adobe Photoshop. The initial graph (a straight diagonal line) has been shifted a little by dragging the black and white end-stops (circled in red) away from the corners. This is a great way to brighten the traces while darkening the background.

Figure 56: *(right)* In this case, the end-stops are untouched; but the curve has been lifted, primarily in the mid-dark part of the image. You might do this to brighten the background a little in, say, a reference image (see page 34). It's important to keep the curve smooth to avoid unwanted effects.

In addition to adjusting the brightness and contrast in this way, you may like to experiment with cropping, rotation, noise reduction and other effects to produce images that meet your objectives. Note that many of the images in this book have been cropped to maximize the clarity and impact, and to enhance what I am communicating about the light sources or their traces.

The art of photographing time-varying light sources

In addition to simply recording how light sources change over time, you may find (as I have) that photography of time-varying light sources approaches an art-form, and I would imagine that readers with an artistic inclination would find a lot of satisfaction in exploring this medium.

The main guidance I would give you for art photography is as follows:

- Look for interesting light sources or groups of light sources.
- Set your exposure time to something longer than you would for documentary pictures. 1/4 second is probably the minimum, with interesting results possible with longer exposures such as one second.
- Move the camera in creative ways, including, for example, zooming while rotating (not as hard as it sounds: grab the zoom ring on your lens and rotate the camera!)

Here are some examples of pictures I have taken which are visually attractive.

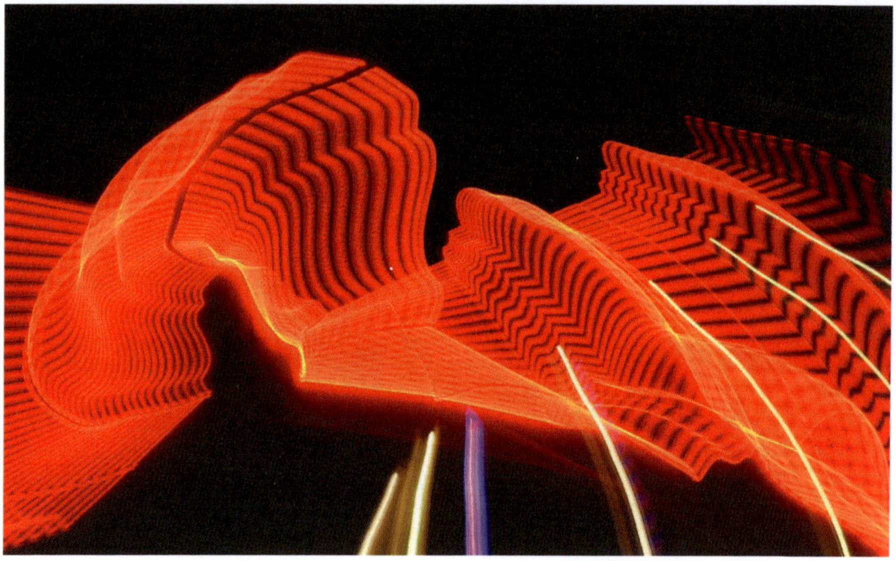

Figure 57: Neon lighting (page 101) makes a great subject for art photography. (0.3 sec)

Figure 58: An example of creative post production. This is the same image as Figure 121, except that here it's "inverted", or turned into a negative. This can be achieved in Adobe Photoshop by pressing "Ctrl" and the "i" key.

Figure 59: A photographic trace of several streetlights (including a neon "vacancy" sign, taken with a simple up-and-down sweep. (1/4 sec)

Figure 60: The same scene, but I used a circular sweep, which produces a different effect. (1/4 sec)

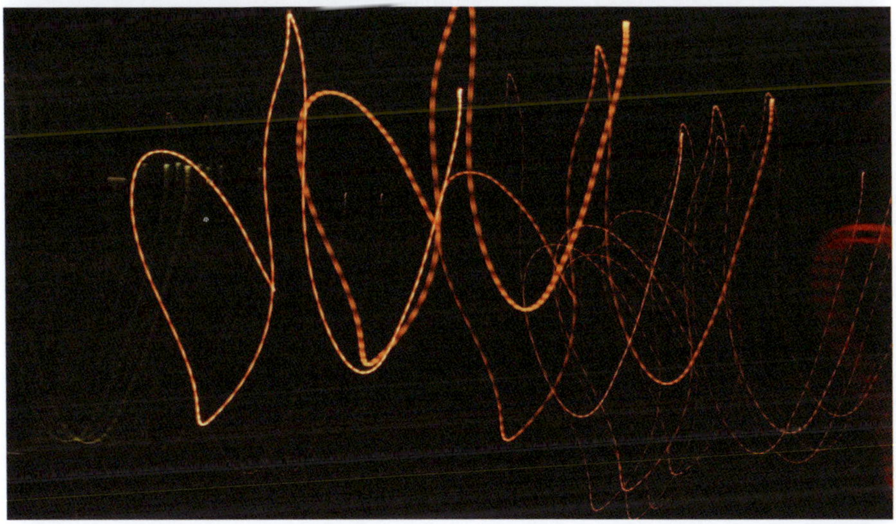

Figure 61: Another shot of the scene, using a longer exposure (1.0 sec) and a more complex motion applied to the camera. This is essentially painting with light sources.

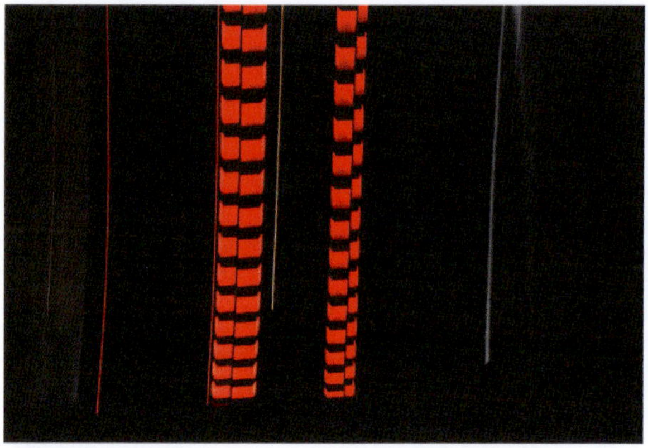

Figure 62: This striking picture is a photographic trace of the tail lights of a 2012 Audi A6. You can learn more about tail lights on page 59. (1/10 sec)

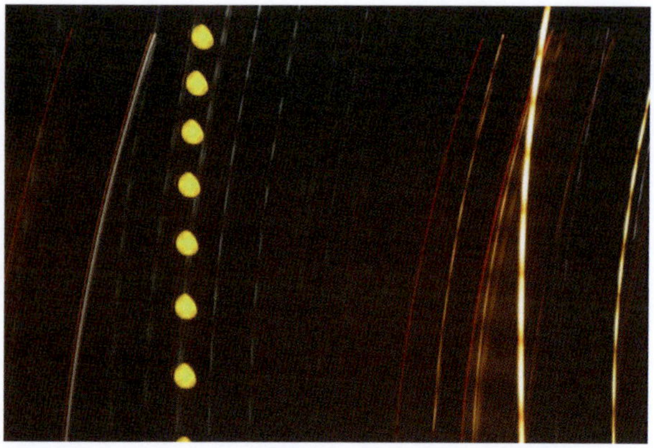

Figure 63: An attractive trace of the yellow light on a road warning sign (see page 66), along with fainter traces of many other lights in the scene. (1/10 sec)

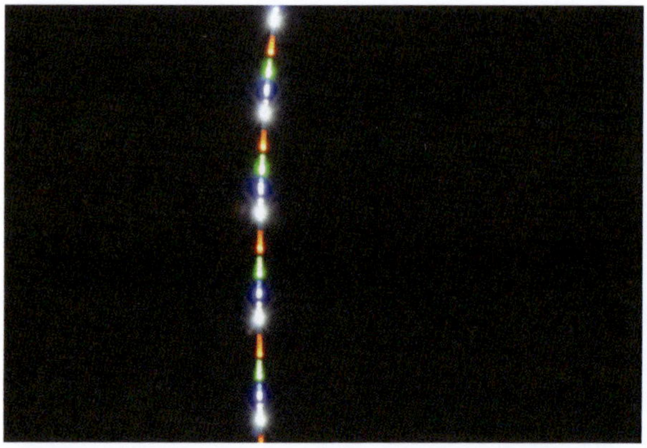

Figure 64: A trace of a digital projector from the front. The image is slightly over-exposed which renders the white flashes very bright. The star-like patterns are due to light scattering on the camera's iris. See page 93 for a discussion of digital projectors. (1/10 sec)

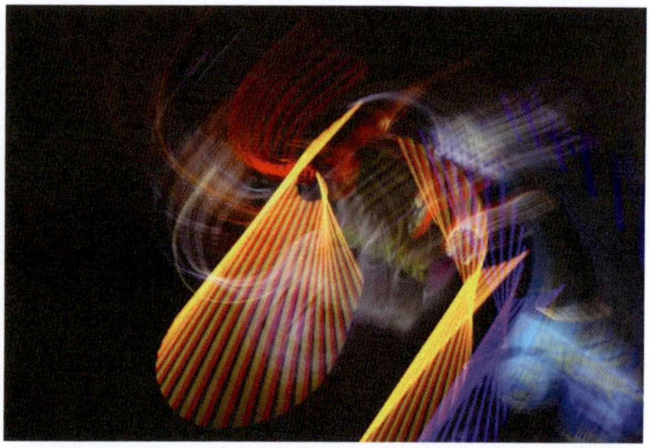

Figure 65: Believe it or not, this is a slot machine at a casino in Las Vegas. I took this picture by rotating the camera in a complex motion while shooting. You can find more photos of Las Vegas on page 127. (1/5 sec)

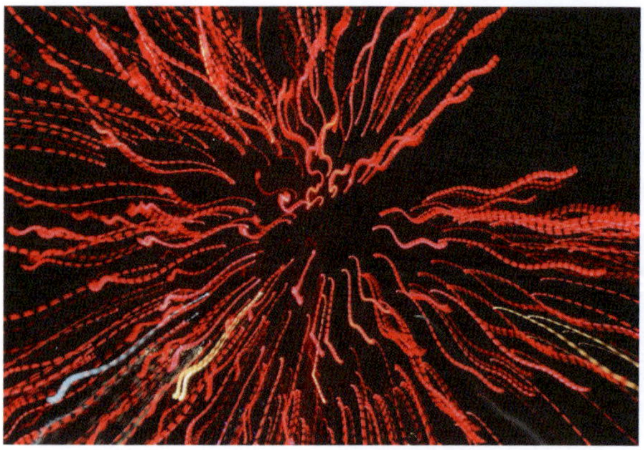

Figure 66: A trace of a tree adorned with red LED Christmas lights. I took this picture by zooming the lens and slightly shaking the camera while taking the shot. You can learn more about Christmas lights (and see more of these images) on page 116. (1/4 sec)

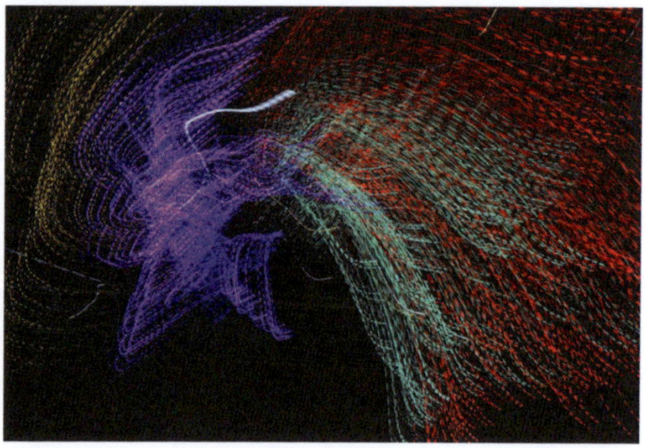

Figure 67: Several trees, each with LED Christmas lights of a different color. I shook and rotated the camera during the shot. Every LED is time-varying and left a dashed trace. (1/4 sec)

Light painting

Those of you who have an interest in photography may recognize that the methods just described amount to a technique called "light painting". This usually consists of placing the camera on a tripod, setting the shutter to be open for an extended time, and then moving one or more lights in front of the camera, literally painting a design with light.

A quick search on the web will show you many astonishing examples of light painting. I have included three here.

The artistic images on the previous three pages would look at home in a gallery of light painted photographs, as that is exactly what they are.

The primary uniqueness of the photographs in this book comes from two things: firstly, that the light sources are time-varying and are the primary point of interest; and secondly that in almost all cases, the camera was most definitely not on a tripod but was moved during the exposure (also a recognized light painting technique, by the way).

Most light painting artists use lights which are constantly on, and which therefore don't really show any variations along their traces. One notable exception is the pixelstick from Bitbanger Labs.

Figure 68: An example of light-painting, where a man is spinning burning steel wool on a rope. The success of this image comes from the fact that the camera's shutter was open for an extended time – 31 seconds, in this case. (Photograph: JJ Harrison)

Figure 69: Another striking image produced by light painting. In this case, the person is twirling an LED-illuminated hoola-hoop. Individual images of the hoola hoop can be seen in different colors, which shows that the overall color of the hoop must have been changing rapidly. The dashed look of the lines arises because there are gaps between the individual LEDs and are not time-varying traces in their own right. (Photograph: Chris Phutully)

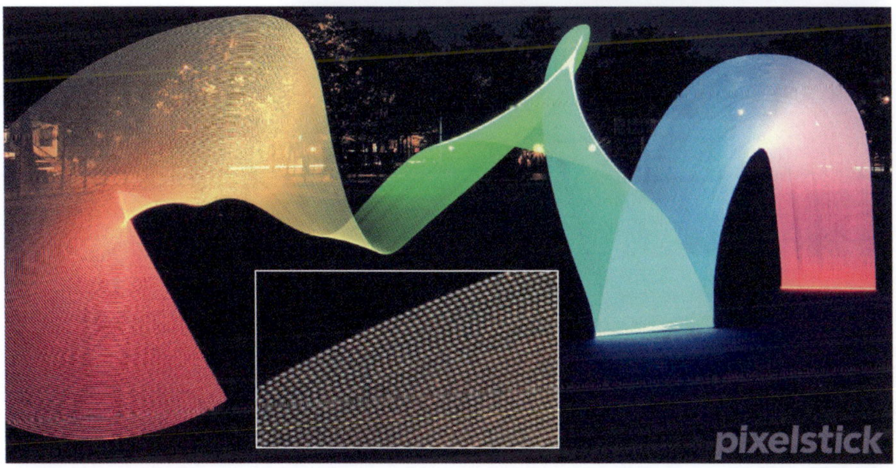

Figure 70: A light-painted photograph generated by loading an image consisting of a horizontal sequence of colors into the pixelstick's memory, then "playing" the image column by column over time as the pixelstick is waved in a flourish in front of the camera. The inset shows a detail of the left part of the trace, revealing discrete dots, showing that the pixelstick uses pulse-width modulation to control the LEDs' brightness (and hence color, by mixing red, green and blue). This modulation is at a very high frequency to avoid it introducing artifacts into the image. Image reproduced by permission. www.thepixelstick.com

Videography

Good and bad

As it happens, I have not spent much time investigating the use of video cameras for observing time-varying light sources, as for a long while I assumed that capturing time-varying light sources with a time-varying device would result in spurious and confusing results.

It turns out on experimentation that I was to some extent correct. It is indeed possible to see traces in the video footage that do not accurately represent what the light source was doing in real life. However, I have also been pleasantly surprised at the results produced by certain cameras, which even have the advantage of making traces visible to people without the need to jiggle one's vision or to use a still camera. In other words, jiggling a video camera can – with care – produce compelling videos of the traces of time-varying light sources.

Just as with vision jiggling, the traces seen in a video are transient, and don't lend themselves to scrutiny and analysis in the way that still images do. It is possible, of course, to extract any single frame of video as a still image, and inspect it. However, remembering that video cameras can introduce unexpected inaccuracies, this is not recommended.

Having said that, I am presenting here a few frames captured from video cameras, to illustrate the strengths and weaknesses of time-varying light source videography. I would love to show you the videos, but in the format of this book it is of course not possible to do so. For that, you should go to the web-site, *timevaryinglights.com*.

Video cameras are time-varying devices

The primary issue when videoing time-varying light sources is that video cameras are themselves time-varying. There are several ways in which this is true:

- Video cameras capture frames at a certain frame rate. Typical frame rates are 24 frames per second (fps), 25 fps, 30 fps, 50 fps and 60 fps. The problem is that if a light source is, say, flashing briefly at 40 fps, then capturing video at 25 fps is not going to reproduce that flash rate, since the camera is not making new frames quickly enough. At other ratios of light source flash rate to video frame rate, you will see mild or severe inaccuracies where the trace has additional or missing parts. Without knowing that this might be happening, you might be tempted to assume that the light's trace is more complex than is really is. You might think that shooting video at higher frame rates, such as 60 fps, would improve matters, and to some extent that's true. However, for a fast light source such as an LED that's flashing 1,000 times per second, 60 fps is still much too slow. And in fact, a *slower* frame rate can sometimes help as each frame becomes a photograph with a substantial length of accurate trace in it.

- The second time-varying issue is that some video devices, such as cellphones, may use a technology called "rolling shutter", which avoids the use of a physical shutter in front of the sensor, but can result in each row of pixels being captured at a different time. This is due to the fact that the image is saved sequentially from top to bottom. Therefore, the lower part of each video frame may actually show the light source at a later time than the upper part. This can potentially introduce distortions in the shape of traces. However, this issue is not as severe as the first issue, and many video cameras do not obviously show this problem.

- Another source of perceived issues is that when watching a video clip you are never seeing one frame in isolation. Instead, you see a sequence of images. Therefore, for example, a frame that accurately captured the flashing light source followed by a frame that did not may give rise to a perception of a complex pattern that wasn't there, and which cannot actually be found in any given single frame.
- For those who know what "shutter angle" is, it should be obvious that this parameter can have a significant effect on the traces you see in the video. (Shutter angle is essentially a way of expressing the shutter speed (i.e. exposure time) as a proportion of the video frame rate. 180° means 50%.)
- Finally, whenever you watch the video you are watching it on a display which is itself a time-varying device! This can add some uncertainty to what you are watching. It is quite common to watch a video of a given frame rate on a display that is refreshing at a different frame rate. In fact, you do this almost every time you watch a video on a computer. This can also introduce artifacts in the traces. I have seen issues where the traces in a video play back perfectly on the camera's built-in display but look broken and inaccurate when played back on a computer. In fact, without the use of a specific "playout monitor" that is synchronized to your video frame rate, it may even be impossible to see smoothly presented traces.

For these reasons and due to my relative inexperience with time-varying light source videography, I am not presenting this topic with the same level of depth as I did for photography. Hopefully the reader will bear with me. You are of course free to... experiment!

Some video frames

Here I present some images which are frames from video cameras. In all cases, the light source is the battery charger discussed on page 26. The important characteristics of the traces it generates are: the two LEDs are on for 50% of the time and off for 50% of the time; and the on and off times are in antiphase: when one is on, the other is off. I selected this source because it will be clear if the video cameras are capturing the traces accurately.

Figure 71: A frame from a video captured on a high-end camera (Canon EOS C100), in 60i format (60 frames per second). This image shows exactly the type of issue you will frequently encounter with videoing time varying light sources. The shutter was open long enough to capture a complete cycle of the 60Hz sequence that the LEDs go through. However, because it happened at a time that's a little offset from the LEDs' cycle, we can see two "dots". If played back at 60 fps on a display that's synchronized, this may not be a problem, because the next frame will continue these truncated traces and the result will look much as it does when vision jiggling. However, as soon as there's any reason for the frames to be played back at another rate there's a strong chance that you will visually see these dots in the video. This may lead you to conclude that the light source has a more complex time-varying behavior than is actually the case.

Figure 72: A frame from a video captured on an HTC Rezound cellphone. Ironically, it's doing a much better job than the C100! The frame rate is slower (22 fps) so the shutter is open for nearly three cycles of the battery charger's LEDs. Therefore, this frame contains a substantial piece of correctly-presented trace. In fact, the traces in the video look better when played than the C100's video.

Figure 73: A frame from a video taken on a Samsung HMX-F80 consumer camcorder. It behaves very much like the Rezound (above). At 30 fps, it captures a little less of the sequence of dashes than does the Rezound. Visually, the video when played looks very good as far as the traces are concerned, closely matching watch you'd see when vision jiggling.

Although it's not easy in a book medium to discuss and present the strengths and weaknesses of digital video as a medium for capturing and viewing time-varying light sources, I hope I have at least indicated that the topic is not as straight-forward as you might expect.

I imagine that the some readers will find interesting and captivating ways to use video in the world of time-varying light sources.

Side-effects of time-varying light sources on videography

Just as we saw with photography, undesirable or unexpected results can occur when capturing video of time-varying light sources. With videography, these side-effects can be particularly severe. The reason for this is that the video camera is itself a time-varying capture device, and if its operation is not explicitly synchronized to variations in the light sources, it is possible to see a "time-domain aliasing" effect, where the apparent behavior of the light source is exaggerated or distorted. We have all seen TV advertisements for cars where the wheels on a forward-moving car *appear* to be turning backwards; this is a familiar example. (Note, however, that many car ads today are created using computer graphics technology, in which case backwards-rotating wheels will have been deliberately animated that way because they look cool!)

Another example is when a CRT TV or display is in the shot. It is common to see dark bars or other variations in brightness apparently moving up or down the screen.

With time-varying light sources such as LEDs, the effect can be even more pronounced. As the camera's shutter opens repeatedly, it may do so at times when a rapidly-flashing light, which would appear "always on" to the human eye, is in its "off" phase. In fact, if the shutter's repeat rate is similar to that of the light source, there can be several consecutive frames during which the light was off, followed by several frames when it was captured during its "on" phase. The result in the output video is a slow on-off flash which does not at all represent the *perceived* "always on" state in reality.

Although I can't show this effect in this book, you can see a video on my website (time-varyinglights.com) on the Videography page. From 1m10s into the video, you will see the LEDs on the control panel of the Korg M3 synthesizer apparently flashing vigorously. Needless to say, the synthesizer's lights do not appear to flash like this in real life.

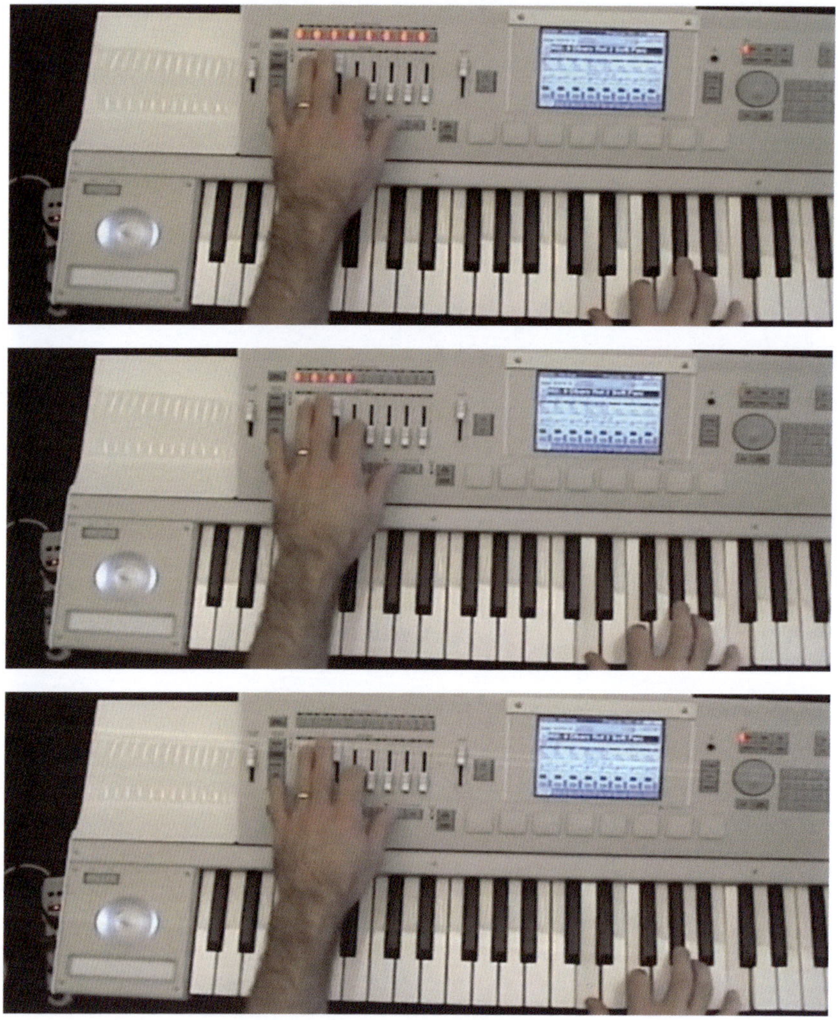

Figure 74: A series of three not-quite-consecutive frames (cropped to fit on this page) from the video referred to in the text, showing the Korg M3 synthesizer. Notice how the red LED indicators on the control panel appear to be on in some of the frames and off in others. This flashing effect is extremely obvious in the video. LED indicators like these are discussed on page 92. Images used with permission: ksounds.com

The Field Guide

This is the start of the Field Guide proper. The remainder of this book is arranged into sections that cover classes of light sources, such as street lights or electronic devices. Within each section are individual types you are likely to encounter as you explore the world of time-varying light sources.

Bear in mind that this Field Guide is not exhaustive; doubtless there are interesting light sources that I have never heard of, let alone seen. And some light sources, such as matrix displays, come in so many varieties that it's impossible to cover all of them.

The main aim of this book is to train you, the inquisitive reader, how to observe time-varying light sources and how to interpret what you see. The Field Guide is designed to launch you on your journey of discovery by showing you what's out there, what to look for, and what you can learn from it. With practice and thought, you will be able to observe and hopefully understand any time-varying light sources, including those that are not described in this book.

Throughout the Field Guide, observation tips (indicated by this symbol:) will guide you to look for important features relating to the light sources under discussion.

A note on the photographs

Although there is a section on photography in this book (page 27), which I recommend you read, it is worth restating here two points that should be borne in mind when you look at the photographs in the Field Guide.

Cameras do not behave the same way as human vision:

- Colors may differ, especially if the light source is bright. Cameras tend to wash out colored lights towards white. The human eye does the same thing, but it takes a much higher intensity to make it happen in the eye.
- When you jiggle a camera, the entire scene is blurred. Human vision, however, is much more complex, and you will find that the background scene is remarkably clearly visible even when you are jiggling at a light source, for example, a street light in front of a sunset.

Street lights

Street lights are among the easiest time-varying light sources to observe. They are ubiquitous and almost all of them are time-varying. A quick night-time trip to the outskirts of a town will provide the best setting for observing street lights, as it can become a little confusing if there are too many lights around you, such as in the center of town.

Incandescent

There are still some street lights which use incandescent light bulbs, especially in less developed parts of the world; but in countries like the USA it is rare to see them these days. They are electrically inefficient, which is why they have been supplanted by discharge lights such as high-pressure sodium and mercury. What's more, they are not interesting to the observer of time-varying light sources (see page 69 to see why), so we will move on to other types without further ado.

Low pressure sodium

Low-pressure sodium lights are rare in the United States, although they can be found in some cities in the Bay Area, such as San José, California. They are very common in Great Britain where they have been the predominant form of street lighting for many years; and they can be found at pedestrian crossings in Germany.

These lights have a distinctive orange or amber color which is due to the characteristic emission spectrum of sodium. Due to the band of emitted wavelengths being extremely narrow, it is almost impossible to tell what color something is when illuminated by this type of lighting. Yellow or orange cars appear white, and red cars appear gray! This is an effect called illuminant metameric failure, in an extreme form. However, since the light's color properties aren't the topic of this particular book, we will leave it there; except to say that if you can find a pair of neodymium-impregnated glasses (used by glass blowers to eliminate sodium-orange glare so that they can see what they are doing), then take them with you and go for a night time walk in English suburbia.

Figure 75: A photograph of a pole supporting four low-pressure sodium lights. This image was exposed to allow the pole to be visible; however, the lights are over-exposed and don't show the correct color.

From a vision jiggler's perspective, these lights are fairly straight-forward: being discharge lights which are powered by both half-cycles of the alternating current (AC) supply, they flash rapidly at 120Hz (in the USA; 100Hz in Europe). Typically, they are illuminated for about 90% of the time, and strikingly dark for the remainder.

Interestingly, some lights follow this pattern but also show a very brief flash right in the middle of the dark period.

Figure 76: A second photograph, this time aggressively under-exposed, which provides a much better representation of the characteristic yellow-amber color of low-pressure sodium lighting.

- 👓 Observe the time-varying behavior of low-pressure sodium street lights. Try to find the type that shows the extra flash in the middle of the dark period.

- 👓 Now look at two or more distant street lights and try to determine how they are connected to the three phases of the AC power supply. This observation is covered in more detail in the next section (Pressurized sodium).

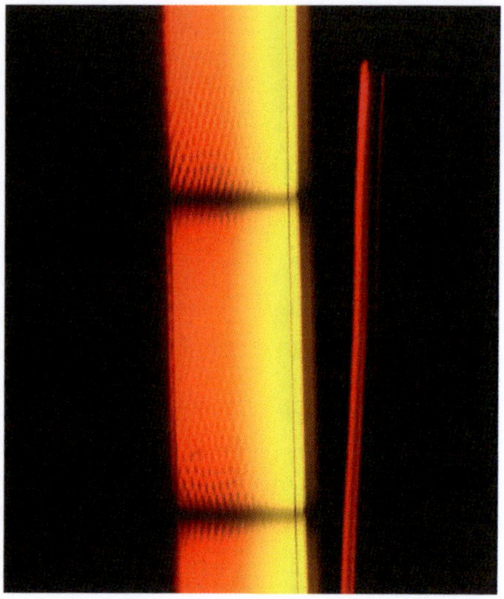

Figure 77: *(left)* A photographic trace of the lights in the previous two figures. Note that the dark bands are very distinct, indicating a sudden cessation and resumption of illumination. Also note that - for these lights at least - there are wavy boundaries which indicate that the electrical discharge is not evenly distributed along the tube at all times. (1/10 sec)

Figure 78: *(right)* A fascinating trace of a low-pressure sodium street light in San José. This lamp was malfunctioning, and showed a red-to-orange gradient to the naked eye. (The red color comes from neon gas, which is also in the tube along with the sodium vapor.) Eyeball jiggling revealed a hint of some complex time-varying behavior, and careful photography confirmed it. You can see a fish-net pattern of light and dark ripples in the tube, which indicates that the gas discharge was experiencing a periodic variation. The diagonal slant of the ripple pattern shows that the ripples were traveling along the tube or forming standing waves while periodically changing phase. (1/10 sec)

Pressurized sodium

Pressurized sodium street-lighting is the predominant type in many countries, including the United States. It has a distinctive apricot color, and emits a broader color spectrum than low-pressure sodium, making it (somewhat) easier to see the colors of objects illuminated by its light.

Because they are discharge lights driven from an alternating current (AC) power supply, they are time-varying. While not the most interesting light source for the vision-jiggler, they are easy to observe. The intensity of the discharge ramps up and down pretty much as the voltage of the supply rises and falls, so the illuminated dashes in the retinal graph will be soft-edged.

👓 Observe a pressurized sodium street-light to see its time-varying behavior. Hint: Don't stand too close to it, or it will be too bright to observe.

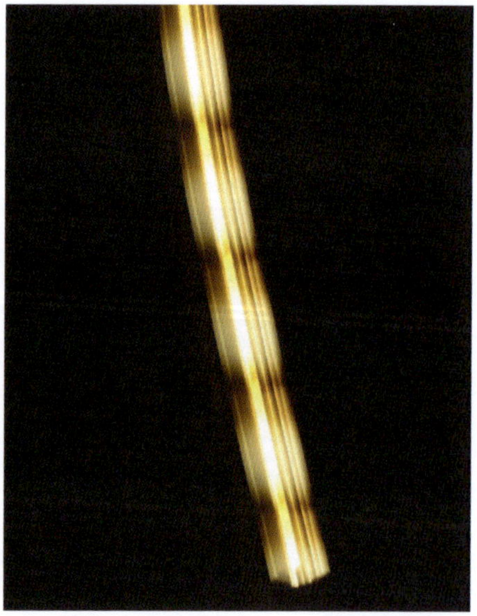

Figure 79: *(left)* A typical pressurized sodium street light. This image is exposed to show the pole and cables, but the light itself is overexposed and has lost its color.

Figure 80: *(right)* A photographic trace of a pressurized sodium light. Notice that although most of the light varies periodically in brightness (and, to some extent, in color), there are some streaks which remain brighter through the dark periods. Many pressurized sodium lamps show this characteristic, though it is not always so obvious. It is attributable to the fact that these lights run at very high temperatures (which is how the high pressure of the sodium vapor is achieved); and the hottest parts of the bulb are hot enough to glow. It is these parts that remain bright. (1/10 sec)

One of the most interesting observations of street-lights (of any type) is to see how individual lights are connected to the three phases of the electrical supply.

Electrical power is distributed in three phases, as this balances the rotational load on the generators in the power station; and it also gives flexibility for electrical installers to easily provide power for low-power tasks (such as domestic lights and outlets) and heavy-duty tasks (such as factory machinery). For street lighting, the result is that each individual street light will be on a particular phase of the electrical supply, and other street lights might be on the same or different phases to it. Although it is not possible to identify *which* phases are used in a row of street-lights, it is certainly possible – with eyeball or mirror jiggling – to observe which lights are on the same phases as each other, and to make some judgments about the electrician's approach when they were wired.

To make this kind of observation, look for two or more street-lights that are visually close together, and are far enough away from you that they are not too bright.

The best results are obtained when you jiggle along an axis that is perpendicular to the imaginary line connecting the lights. See the following figures for an explanation.

Figure 81: Here are two pressurized sodium lights outside a building. Note that they are approximately at the same height in the photograph, or in other words, they are horizontally aligned. Now consider the following figure.

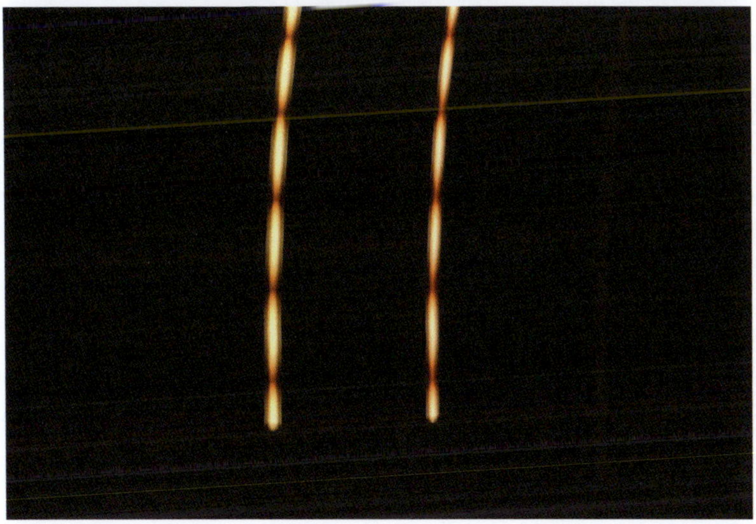

Figure 82: Here is a photographic trace of the lights in the previous figure. The camera was swept vertically, which is *perpendicular to the line connecting the lights*. It is possible to see that the dark periods of the two lights are "side by side". This shows that the lights are flashing at the same time, which indicates conclusively that they are both connected to the *same* phase of the electrical supply. (1/10 sec)

What you are looking for is a certain relationship between the locations of the features in the traces. If they line up, the lights are connected to the same phase; if the features are offset ("by a third") then they are on different phases.

If you jiggle in the wrong direction, there is a risk that you will misinterpret the phase relationship because the "starting point" of one light is shifted, as described in the following figures.

Figure 83: Consider the two pressurized sodium street-lights in the center of this photograph. They are not aligned horizontally, so if we were to sweep the camera vertically, the traces will be offset from each other simply by virtue of the spatial positions of the lights.

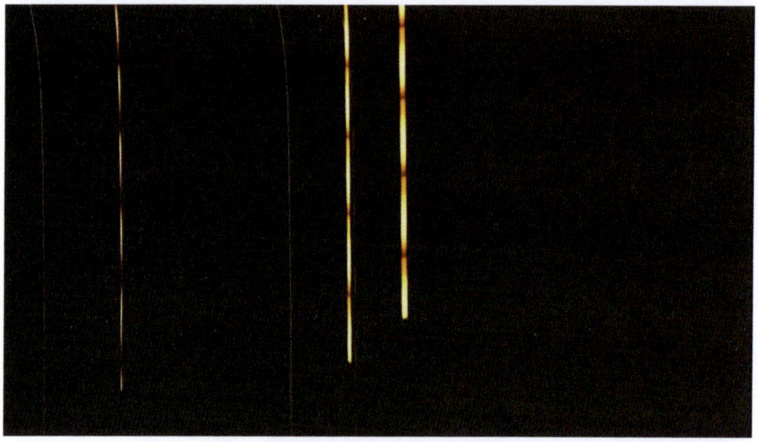

Figure 84: On inspecting a downwards-swept photographic trace of the lights in the previous figure, we see that the dark periods in the traces do *not* line up. This might lead us to the conclusion that the two lights are on different phases; however, we need to account for the specific arrangement of the lights in the scene. Since the left light is lower than the right one, it's reasonable to expect that its *trace* will also be lower, leading to an offset that might confuse an attempt to determine if the lights are on the same or different phases. To be fair, in this photograph it is possible to determine the answer, because we can see the *starting points* of both traces, and it's clear that the patterns are identical, so it's safe to say that the lights are indeed on the same phase. However, when vision-jiggling you will not be able to see the starting points of the traces. In a photograph, the starting points simply correspond to the moment that the shutter opened. (1/10 sec)

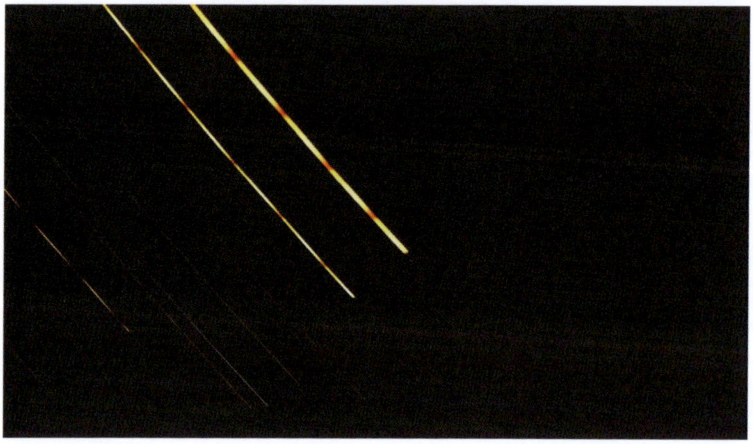

Figure 85: For this photographic trace, the camera was swept along a line that is *perpendicular to the imaginary line which joins the two light sources*. This essentially compensates for the spatial offset, and causes the traces to line up. Now it's possible to determine if the lights are on the same phase of the electrical supply or not; in this case, they are. To achieve this with eyeball jiggling, you will probably need to tilt your head so that the direction of jiggle is perpendicular to the line connecting the lights. With a mirror, jiggle it along that axis. (1/10 sec)

With the technique perfected, let us now consider what you will see.

Start with just two lights. If they are connected to the same phase of the electrical supply, the features in their traces will line up as in Figure 82 earlier in this section.

If they are on different phases, their traces will misalign. Figure 87 shows both in-phase and out-of-phase street lights so you can see the difference.

Importantly, the misalignment will be by exactly one third. This is because the three phases are equally spaced in time at one-third intervals. Electrical engineers call this a phase difference of 120°, since 120° is a third of 360°. (Strictly speaking, the misalignment can also be by two thirds; but this will also look like a one third displacement, just in the opposite direction.)

Now let us observe a whole row of street-lights.

Again, due to the consideration of start-points, this observation works best if the lights are exactly in a line with each other, and if you jiggle along the direction which is perpendicular to the line of the lights. In this case, you will see a whole array of traces, and – if you observe carefully – you should be able to identify patterns in the phase relationships. You may find it helpful to photograph the scene and study the photographic trace, such as the one below.

Figure 86: A row of streetlights, most of which are of the pressurized sodium type.

Figure 87: A photographic trace of the scene in the previous figure. Eight of the pressurized sodium lamps are prominent. Inspection shows that all are in phase except for the two which are immediately to the left of the dim blue vertical line. Note how the dark bands of these two lamps are displaced from those of the other six by a distance of one third. (1/10 sec)

In summary, you can tell at a glance (or with a mirror) whether two street-lights are connected to the same phase of the electrical supply or not; and with a row of lights, you can tell whether the electricians connected them in any regular pattern to the three phases. This will amaze your friends; but it may not be something you want to demonstrate on a first date...

Since a pressurized sodium street light is strobing (flashing rapidly), it follows that the same will be true for all objects illuminated by these lights. Find a white line painted on the road under a street-light, and jiggle at it. You should see apparent time-varying behavior. A parking lot is a good place to experiment.

Figure 88: Stop lights on a gantry which also carries a pressurized sodium street light (at the top of the photograph).

Figure 89: A photographic trace of the structure in the previous figure (the stop lights have changed to green). Note that the gantry, which is illuminated by the pressurized sodium lamp, appears to be varying in brightness, leaving a trace of faint horizontal bars. Note that the motion of the camera was not uniform, which is why there is a fairly clear image of the scene in addition to the swept traces. (1/10 sec)

An extension of this is to observe something that is moving. Try having a friend ride a bicycle while wearing a white T-shirt in a parking lot under pressurized sodium lighting: you may see some time-varying effects. Alternatively, wave a white stick rapidly under the street light.

It is also interesting to watch raindrops as they fall near a street–light of this type:

observe "dashed lines" of light as each drop is illuminated and then not illuminated by the lamp as it falls. Note, however, that raindrops can sometimes show dashed-line behavior even when the illumination is *not* time varying; page 147.

Figure 90: Rain falling past a pressurized sodium street light. The bright top part of the light is out of the picture. The faint smudge near the bottom of the picture is "lens flare", which arises from reflections of bright light sources in the lens. Note how the raindrops leave dashed patterns. This is a result of the time-varying illumination from the light. Since the camera's shutter was open for 1/10 second, each raindrop should show twelve bright dashes (the light is flashing at 120 Hz). Raindrops that have longer dashes were falling faster than ones with shorter dashes. On close inspection you can see a few traces which arc away from the light. These are the result of rain splashing on the light fixture. Notice that these drops were moving more slowly: they hadn't fallen all the way from the clouds above. (1/10 sec)

All of these observations apply to almost all other types of strobing street-light, such as mercury discharge or low-pressure sodium.

Mercury discharge

Mercury discharge lights are also common in many places. Here in the United States, while there are mercury street-lights, you are also likely to find them on the corners of houses or the front of barns. They are a good choice for security lights for farms and yards.

Figure 91: A mercury discharge lamp mounted on a wooden pole.

For an observer of time-varying light sources, mercury lamps are very similar to pressurized sodium lamps, except that the color is different, being white (or apparently slightly greenish when seen from a distance, at least on first glance).

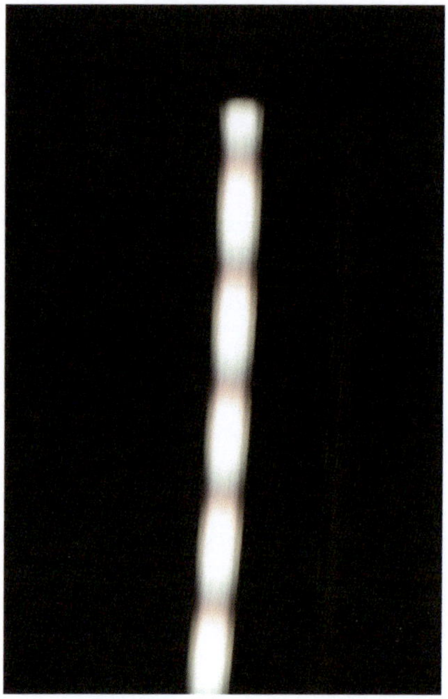

Figure 92: A photographic trace of the mercury lamp pictured in the previous figure. Notice a slight pink coloration at one end of the bright white phase, and slight green coloration at the other end. From this photograph, it's not possible to determine whether the green color occurs (temporally speaking) at the start or end of the white phase. Also notice that the trace becomes broader and narrower as the intensity rises and falls. In this respect, the camera behaves in a similar fashion to the human eye. (1/20 sec)

- Since they are so similar, I refer you to the observations section for pressurized sodium lighting (page 50).
- The most significant difference between this type of lighting and pressurized sodium is the possible appearance of a pink or even maroon coloration to the darkest part of the trace. This can be quite pronounced. And as the photographic trace in Figure 92 shows, there may even be a green period too, though this may be hard to discern when vision-jiggling.

Light Emitting Diodes (LED)

Although not yet common, white LED street lighting is making an appearance, usually in parking lots, gas stations and commercial campuses. I have not yet seen it deployed along public streets or highways, but I expect it will become common in the future.

To obtain the brightest light from each street light, the LEDs should be driven constantly, with DC (direct current) electricity. Many LED street lights are indeed driven this way, and show no time-varying behavior: a quick observation should confirm this. Some LED street lights, however, are modulated, probably to reduce the intensity and save energy. These show clearly dashed trails when observed or photographed.

Figure 93: A typical location for LED street lighting: a recently refurbished car dealership.

Figure 94: One of the lights in the car dealership shown in the previous figure. This light is twin-headed, with each head consisting of an array of modules, each of which has several high intensity white LEDs. The pole can be seen to the right.

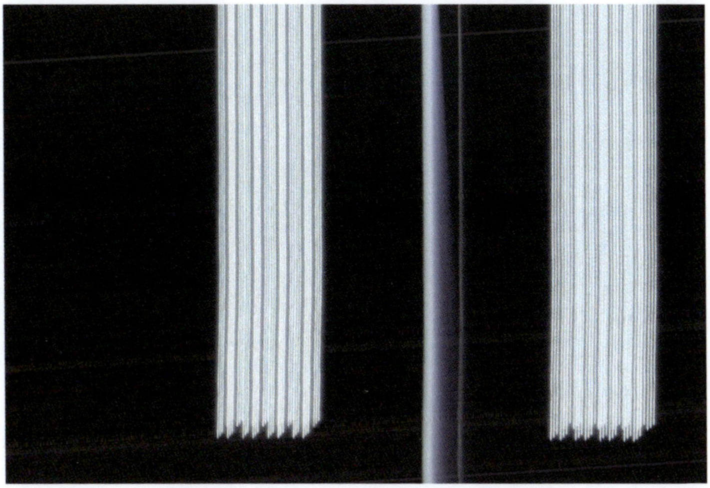

Figure 95: A photographic trace of a pair of lights in the car dealership lot. Two things are apparent: firstly, you can see individual traces from each LED. Secondly, these traces are unbroken white lines: the LEDS are not time-varying. These lights are DC-powered. The white column in the center comes from light reflected from the pole. (1/10 sec)

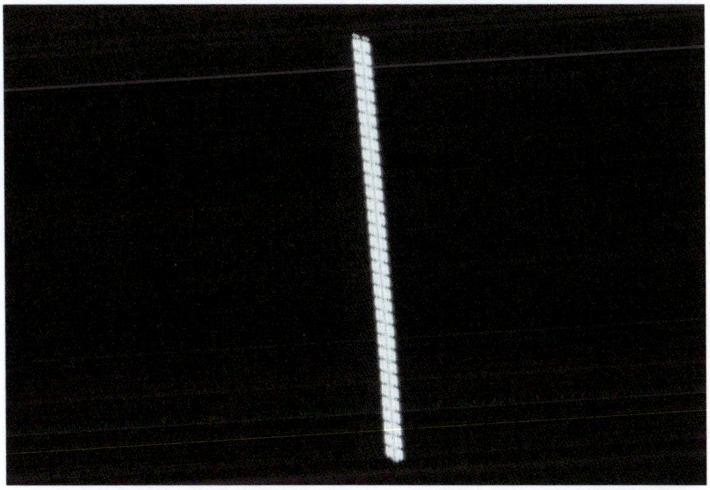

Figure 96: In contrast, here is a trace of an LED street light that is time-varying. Physically, the module that carries the LEDs is much smaller than those seen in the previous figures. What's amazing is the rate at which this light is flashing. The exposure time was 1/60 of a second, and I have counted 34 flashes in this trace. That comes out at 2,040 flashes per second! Even at this rate, it is easy to observe with vision jiggling.

Figure 97: The light shown in the previous figure is one of many installed at a municipal waste-water treatment facility in our local city. Here is a reference photograph showing part of the campus at night.

Figure 98: Here is a photographic trace taken from the scene in the previous figure. The dashed-line nature of the traces is very evident, and this is exactly what one can see with vision jiggling at this location. Note that the ends of each dash are very abrupt, with sudden transitions between fully bright and fully dark. This is consistent with LEDs being driven by an electronic circuit, probably with a power transistor providing the output. In my opinion, this image does not look like it was produced with a camera.

Automotive lights

As I write this book in 2015, most headlights and tail-lights on vehicles are still the traditional incandescent (light-bulb) type. These are not time-varying, and are therefore not interesting in the context of this book. However, many automotive manufacturers are now equipping their vehicles with tail-lights composed of light-emitting diodes (LEDs), which are often flashing on and off very rapidly. These make an ideal target for the observer of time-varying light sources. Some cars also have white LEDs serving as daytime running lights at the front of the car. These too are usually strobed.

Interesting observations are not restricted to car tail-lights, however: stop lights at intersections, dashboards, lights on emergency vehicles and the LED lights used by cyclists can also show time-varying behaviors.

LED tail-lights

While most vehicles are still (as of 2015) using conventional light bulbs in their tail-lights, an increasing number of manufacturers are using light-emitting diodes (LEDs). Current cars from Audi and Cadillac are good examples.

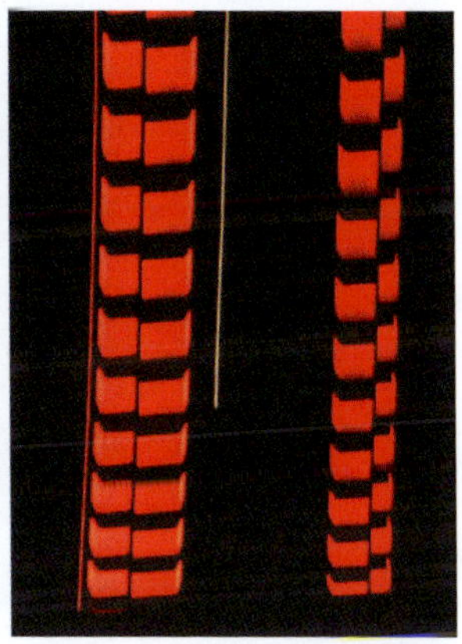

Figure 99: *(left)* A 2012 Audi A6 on a cold night at Ed Carroll Audi in Fort Collins, Colorado.

Figure 100: *(right)* A photographic trace of the A6. The shape of the tail-lights is clearly visible in this trace, which underlines the "light source as paintbrush" concept discussed in the earlier part of this book. You can also see that the shapes are stretched vertically: this is due to the fact that the lights were illuminated for a significant amount of time during each "on" period. This emphasizes the fact that a trace, whether retinal or photographic, is a combination of the shape and arrangement of the light source *and* its time-varying behavior. You can also see that the images of the four lights appear to be a little displaced relative to where they are on the vehicle. This is due to their flashing at different times relative to each other. (1/10 sec)

The thing that makes LEDs especially interesting for the observer of time-varying light sources is that it is generally not possible to change the brightness of an LED by varying the electric voltage or current that is driving it. LEDs broadly tend to be on or off, with very little nuance in between. And yet, vehicle tail-lights are required to deliver two levels of brightness: dim for tail-lights, and bright for brake lights.

The inability of LEDs to offer a range of intensities means that engineers have found another solution to the problem of delivering two levels of brightness: modulation. And modulation is time-varying. Let me explain.

To achieve maximum brightness, which is needed for brake lights, the LEDs are usually driven with a constant current. In other words, they are always (or almost always) on. When designing the car, engineers select high intensity LEDs which serve well as brake lights when driven in this way.

To achieve the reduced level of brightness needed for the tail-lights function, the LEDs are switched on and off rapidly using an electronic circuit. This is modulation, since the brightness is being changed (modulated) rapidly. In fact, they are off for a lot more time than they are on. Typically, they may be switched on for only 10% of the time, then off for 90%, and then the cycle repeats. This happens so fast that the human eye can't normally see it; instead, it *perceives* the light as being much dimmer than when operating as a brake light.

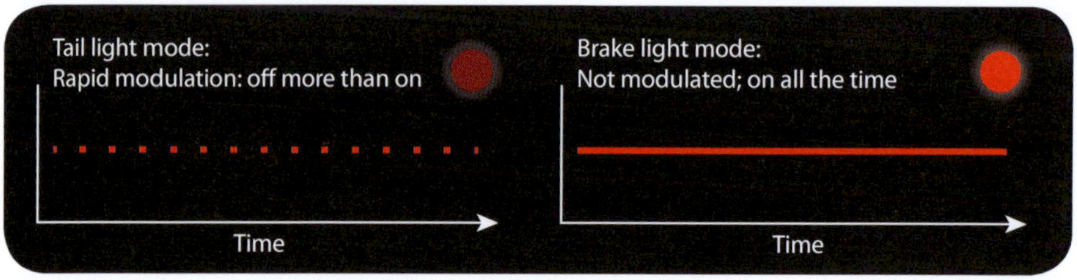

Figure 101: A diagram showing how time-domain modulation affects the perceived intensity of an LED-based tail light.

However, with vision jiggling, it's very easy to see what's going on.

In fact, you can identify whether a car ahead of you has modulated LED tail-lights, even up to a mile away! (I have verified this.) It is so clear that even if you are not actively jiggling your vision you may notice it as your gaze moves while you are driving. I personally find it a little distracting, and I am not a fan of LED tail-lights for this reason.

Bear in mind that each bright image in your retinal trace will have the shape of the tail light itself. For example, on a GMC Acadia, the tail-lights are ring-shaped, so you will see a trail of little red rings.

Figure 102: A GMC Acadia vehicle with the tail-lights illuminated. The camera has over-exposed the red color a little to pink, and the LEDs themselves to white. In reality, they are red, as you can see in the next figure.

Figure 103: A photograph of the tail lights of a GMC Acadia. By under-exposing the photograph, the color of the LEDs is captured more faithfully.

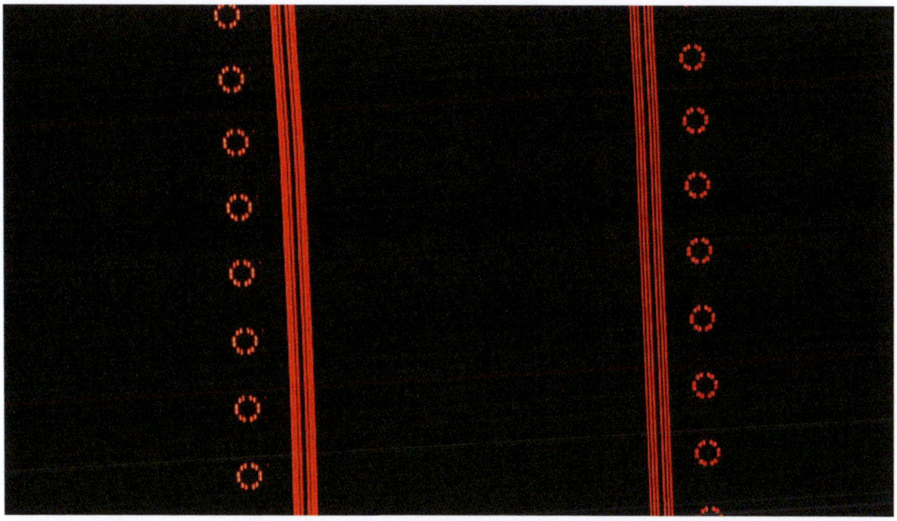

Figure 104: A photographic trace of the Acadia's tail lights. Interestingly, GM's engineers chose to use low-intensity red LEDs for the inner pair of clusters. These are DC powered and hence always on, which gives rise to the uninterrupted red streaks. The outer clusters, however, employ *high*-intensity LEDs and are modulated to deliver approximately the same perceived intensity as the DC-powered clusters. When the driver applies the brakes, these outer clusters become DC powered and are much brighter, serving as brake lights. Notice that the circles are hardly stretched or smeared vertically, which indicates that the "on" time is very brief indeed. (1/10 sec)

- The first observation is to spot vehicles which have strobed LED tail-lights. This is one of the easiest vision-jiggling observations to make: these lights are very obvious, and will appear as a trail of sharply-defined, bright images of the tail lights. They are sharply defined for two reasons: (a) the LEDs turn on and off instantaneously, so there is no ramp-up and ramp-down period, and (b) they are on for just a short time, so there is a large amount of separation between the images.

- Secondly, see if you can learn something about the electronic circuit – or circuits – which control the LEDs.

I have seen some very interesting things relating to the electronic controllers for tail-light LEDs. For example, on a 2012 Audi A8, each tail-light consists of two parts (one on the car fender and one on the trunk), and these two parts are controlled by separate circuits. This type of implementation can be confirmed by noticing that the trails of images are often not lined up in your retinal graph. However, what's even more interesting is that if you observe again a few seconds later, the relationship between the trails may have changed. This shows that the two circuits are flashing their respective LEDs at slightly different rates, and are not synchronized with each other.

In fact, I expect that the control circuit is integral to the light cluster, which means that

Figure 105: A 2012 Audi A8, with tail lights on and brake lights off. Note that the camera has lost the red color of the bright parts of the tail lights. This is typical (see page 33 for more about this effect).

each light cluster has a separate circuit. This would explain the lack of synchronization between the clusters. For the required function – modulating the LEDs between tail-light and brake-light functions – this design works as well as any other. What's satisfying is that just by jiggling your vision, you can see how Audi's engineers implemented the electronics on this vehicle. I have seen vehicles where all the tail-light clusters *are* synchronized, which implies a different design approach using a single master circuit.

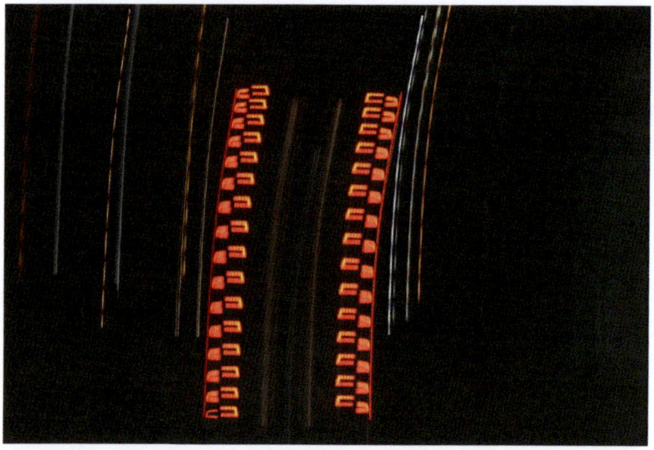

Figure 106: A photographic trace of the tail lights of the Audi A8. Note that the shapes of the red images correspond to the shapes of the tail-lights. Unlike the Audi A6, these images are not significantly stretched vertically, which means that the LEDs are illuminated for a very short time. (1/10 sec)

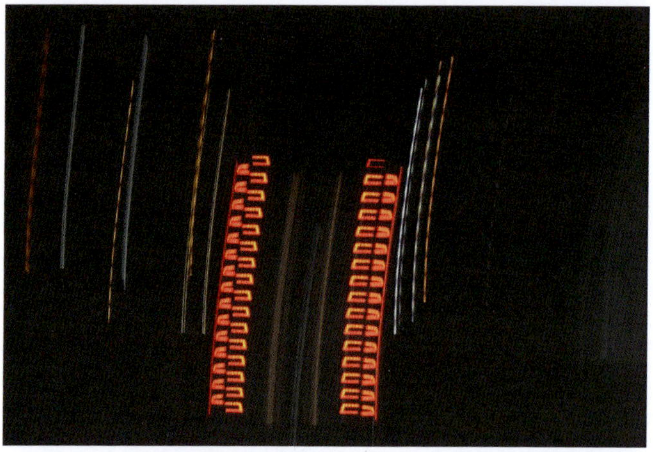

Figure 107: Another photographic trace of the A8. Notice how the *relative* positions of the red shapes have changed from the previous photo. This proves that the circuits that are driving the LEDs are not synchronized, and are flashing at slightly different rates, so the alignment changes over time. The fact that the images appear closer together in this trace than they do in the previous one is simply due to the fact that the camera was swept downwards less quickly for this photograph. (1/10 sec)

Figure 108: A trace of the A8 with brakes applied. Interestingly, the light clusters on the fenders are now DC-powered (they're on continuously) whereas the light clusters on the trunk are still strobed (on most of the time, with regular interruptions). I have no idea *why* Audi implemented the electronics in this way, but it's interesting to observe this quirk. The additional red band in the middle of the photo comes from the line of (much dimmer) red LEDs in the rear window, which comprise the high-level brake light. Note that once again the camera has lost the original red color of the brake lights, instead recording a bright amber color; I should have reduced the aperture size a little (see page 33 for discussion). (1/10 sec)

Finally, look to see what happens when the driver applies the brakes. Are the LEDs, which are now apparently brighter, on for 100% of the time? Or are there short moments of darkness in the trail, which would suggest a pattern such as on for 95% of the time and off for 5%? I have seen both implementations.

LED daytime running lights

It seems that curved strips of white LEDs are a must-have design feature for modern cars. These are frequently modulated in the same way that many LED tail lights are modulated to control the apparent intensity. The result is that it is often very easy to observe their time-varying behavior.

Figure 109: The front view of a 2012 Tesla Model S at a trade show. The white LED running lights which wrap around the headlights are illuminated. Note too that there are two small white side-lights lower down.

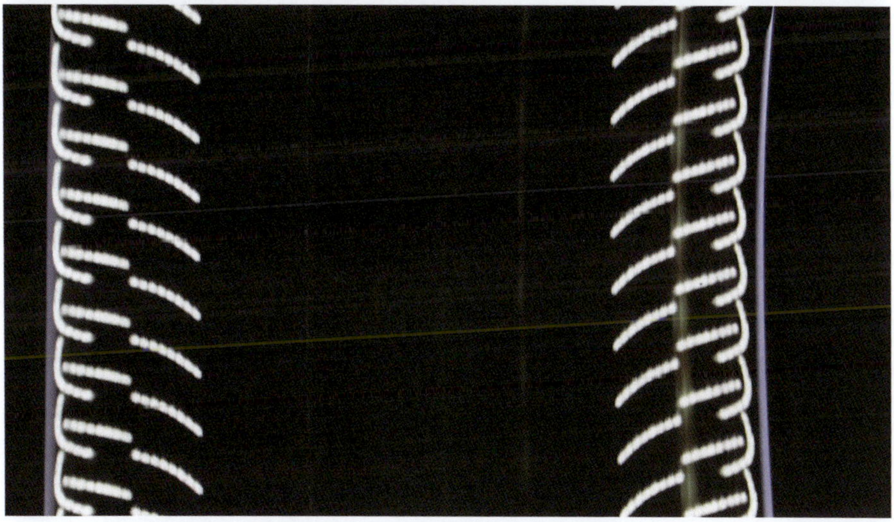

Figure 110: A swept trace of the Tesla Model S. The LED running lights are clearly shown to be time-varying, with marked flashes of illumination separated by periods of darkness. Notice that the strips are actually divided into three sections: one on the side of the headlight, and two along the top of the headlight. These sections are flashed at different times, which is why the images of the running lights do not have the same shape as the physical features on the car. Additionally, note the dimmer blue-white traces that lie just outside the running lights: these come from the white side-lights, which are DC powered with no time-varying behavior. The blue tint arises from the behavior of the sensor in the camera: these lights appear white to the eye. (1/10 sec)

At night, of course, the headlights are also illuminated. These are almost always incandescent and therefore not time-varying; and the very few LED headlights on the market are also not time-varying because they are driven with constant DC power for maximum brightness. I expect that over the next few years, more vehicles will appear on the market with white LED headlights. Note, however, that even with the headlights turned on it is still possible to observe the flashing of the LED running lights, especially with incandescent headlights as these are a markedly different color to the LEDs. With the headlights on, expect to see a solid, unbroken trace from the headlight, overlaid with the periodic flashes of the LED running lights.

Figure 111: A photographic trace of a car's headlights. You can see the unbroken white traces of the incandescent headlights, and the clear repeated images of the LED running lights. In this case, all the LEDs in each light are flashed simultaneously.

Dashboard illumination

There are several forms of lighting used for the instruments on the dashboard of a car. Incandescent lighting – in other words, little light-bulbs – used to be the most common type; but fluorescent lighting and LED lighting have become more prevalent. I used to drive a Lexus RX300 which had fluorescent dashboard illumination.

Incandescent lighting is not time-varying, and so it offers nothing for the vision-jiggler to observe. Fluorescent and LED lighting, however, are time-varying and so you should be able to see a series of images of the dashboard.

Figure 112: The dashboard of a 2011 Kia Sportage.

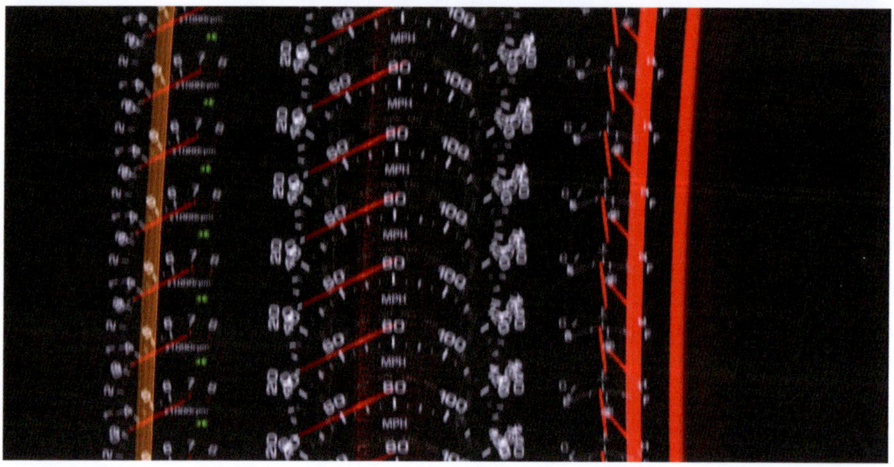

Figure 113: A trace of the dashboard shown in the previous figure. This dash is lit by white LEDs which are modulated to control the brightness. As with most modulated LEDs, the transitions between off and on states are abrupt, which is why the separate images of the dials are so clearly defined. The solid bands of orange and red color come from the warning lights, which are DC-powered. (1/10 sec)

Finally, some vehicles use electroluminescent surfaces for dashboard lighting. These tend to be used for the main dials (speedometer and tachometer), where the whole dial will light up pale green, with the numbers appearing dark. This kind of lighting may or may not be time-varying.

- Determine if the dashboard of your car or truck is illuminated by time-varying or non-time-varying lamps. Simply eyeball- or mirror-jiggle at it in the dark to see if there is a series of images in your retinal trace.

- If your dashboard is illuminated by fluorescent lamps or LEDs, see what happens when you adjust the brightness of the dashboard. Does the intensity of the "on" periods really diminish? Or do they become shorter?

Sometimes you might have LEDs or other display elements on your dashboard which may be time-varying. A 1999 Saab 9^3 uses light bulbs for the dashboard; but there is one LED (which indicates the position of the automatic transmission stick) which is modulated at about a 50% on/off cycle, presumably to tame its apparent brightness.

You may also see numeric displays in your car, such as a digital clock or the display on a car radio. If these are LED or vacuum fluorescent displays, see page 79 for more details. If these displays are backlit LCD displays, they are usually not time-varying.

Stop lights at intersections

Stop lights are interesting due to the variety of types. Even at one intersection you may see variations in type and color. I have often seen three green lights on a single gantry, where one might be incandescent and the others are LED-based. The green color can be quite different between them.

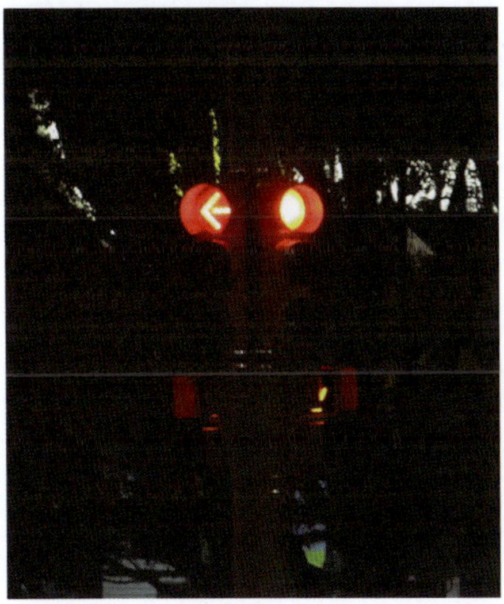

Figure 114: *(left)* Stop lights at an intersection. Notice, by the way, that the digital camera has struggled to capture the red color of the lights. This is typical behavior.

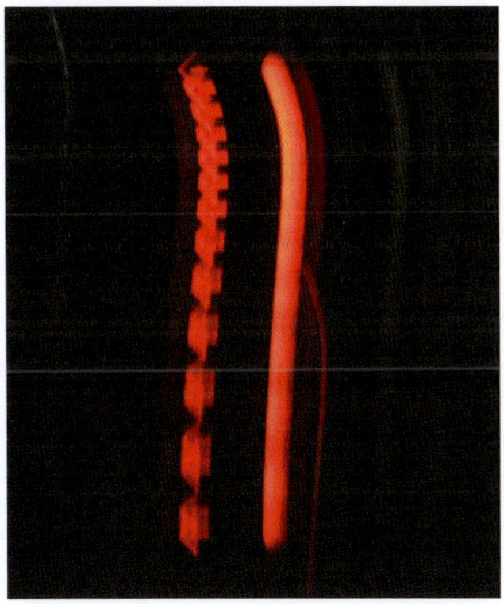

Figure 115: *(right)* A photographic trace of the stop lights in Figure 114. The right-hand (round) light shows only the subtlest time-varying behavior and is probably incandescent. The left-hand light is clearly strobing, and uses LEDs for illumination. There are twelve flashes in this 0.1-second exposure, so the light is flashing 120 times per second (120Hz). The eagle-eyed among you may notice *thirteen* flashes, where the 13th is the thin image at the very top of the trace. However, allow me to point out that the *bottom* image of the arrow is actually *narrower* than the next image by the thickness of that 13th image. This is because the camera's shutter opened a little after the start of an "on" period, so the first arrow didn't paint as long a trace as the other arrows; and the shutter closed a tenth of a second later, just after the start of the 13th flash. So the tenth of a second exposure really is equivalent to *twelve* flashes, not thirteen. This image also illustrates that the camera does a better job of capturing the correct color when the camera is swept.

It turns out that even the incandescent lights can be very slightly time-varying (Figure 123 shows this). But LED lights are the most interesting, because they are always time-varying due to simple rectification of the AC power supply (diodes can only pass current in one direction, not both).

- Determine which, if any, of the stop lights ahead of you are time-varying, and see if you can correlate it to type (incandescent, LED). In our neighborhood, it seems to depend on which city you are in. Fort Collins has a healthy number of LED-based stop lights; Loveland appears to have none.

If there are more than one time-varying stop light on a single gantry, see if you can verify that they are strobing in phase. They should be, since it is unlikely that more than one phase of the AC power supply is used by the system. Refer to the observations for pressurized sodium street lights on page 51 for details on observing electrical phase relationships.

Warning sign lights

In the USA it's common to see a yellow or red light mounted above a warning sign or a STOP sign, to catch the attention of drivers. These lights usually flash on and off slowly, often a little slower than once a second. This flashing isn't interesting for the vision jiggler. What's more interesting is what the light is doing when it's "on". Many of these lights are strobed quite rapidly (though some are incandescent, and not strobing).

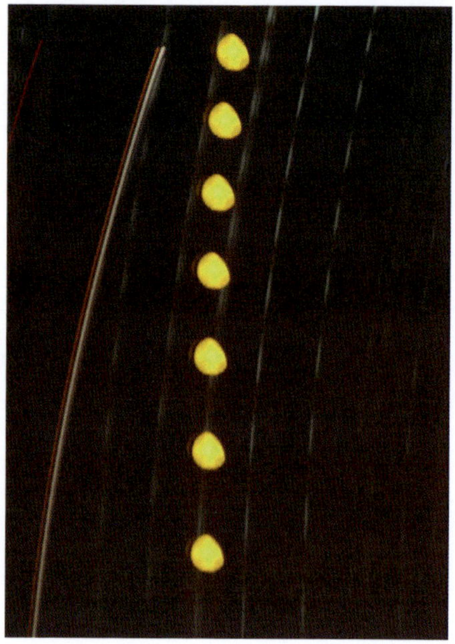

Figure 116: *(left)* A permanently-placed road warning sign, with a flashing yellow light mounted on top. Note that the digital camera has (typically) failed to capture the yellow color of the lamp due to its brightness. The horizontal red and white streaks are from cars passing along a road.

Figure 117: *(right)* A photographic trace of the same light, showing that when the light is "on", it is in fact strobing. Notice that each image is an accurate picture of the light with no vertical stretching. This shows that the flashes are very short indeed, and that the transitions between on and off are instantaneous. In fact, this light is dark for most of the time, even when "on". Sweeping the camera faster, to separate the images more widely, would verify this. (1/10 sec)

Road work warning lights

In road work zones you will often find yellow flashing lights which are typically mounted on top of barrels or cones. These lights may appear permanently on, or like the warning sign lights described above, they may flash slowly on and off. During the "on" periods, these lights are usually strobed extremely rapidly and are a very easy target for the vision jiggler. I expect that the high frequency of *very* short duration flashes is a consequence of the need to make the battery last for a long time.

Bicycle lights

LEDs (light emitting diodes) have almost completely superseded the humble light bulb for bicycle lighting. Nowadays, it's as common a sight to see lights strapped to heads, helmets and backpacks as it is to see them attached to the bicycle itself. LEDs have many advantages over incandescent lights for this purpose: they are brighter, they consume less power (batteries last longer), and they are more robust.

For the vision jiggler, many bicycle lights that use LEDs are interesting because they are strobed. That is, they are switched on and off very rapidly, as a means of limiting the brightness and enhancing battery life. More details can be found on page 59.

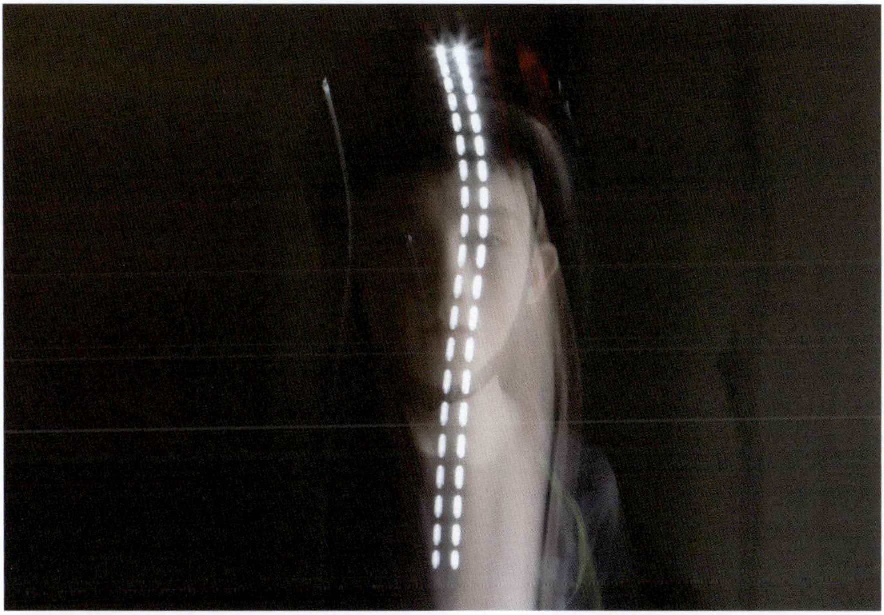

Figure 118: A swept photograph of a boy wearing a cycling helmet with a white LED headlamp. The headlamp has two LEDs, which are strobed rapidly (about 220 times per second). Notice that the two LEDs are always in phase: one circuit controls both of them. The clear appearance of the boy's face is due to the camera happening to be still for a moment at the end of the sweep, before the shutter closed. (1/10 sec)

- When you see lights on a bicycle or the cyclist, determine whether they are strobed (flashing) or steady-state (DC powered). Both types are out there.

- Some cycling lights flash randomly to draw attention to the cyclist. For a light that does this, see if you can observe the high-frequency time-varying behavior of the various types of flashing. This may be hard to do.

Emergency vehicles

It may seem intuitive that the flashing lights on emergency vehicles would present a good target for observation and photography, but this turns out not always to be true. This is because the flashing that you can see is much slower than the type of flashing that a vision jiggler is interested in. Any lights that are comprised of a reflector that rotates around an incandescent bulb are not interesting here.

Figure 119: A reference photograph of a group of emergency vehicles seen one night in our neighborhood in Colorado. The swirls of color are a form of lens flare arising from reflections in the lens.

Modern emergency vehicle lights are based on LEDs or discharge tubes (similar to camera flash guns). These are a better bet for interesting observations. However, bear in mind that emergency vehicle lights are designed to catch drivers' attention, and are therefore visually changing all the time. This can make observation a little challenging, as what you're looking at doesn't stay constant for long. In general, a quick jiggle can identify whether something interesting at high frequency is happening, but then one should turn to a camera for images that can be enjoyed and perhaps analyzed.

Figure 120: A photographic trace of the scene in the previous image, which broadly lines up so that it's possible to identify which lights are causing the traces. Note that some lights, such as the bright red ones in the rear window of the left-most vehicle, were dark when the reference photo was taken, but illuminated during this shot. This explains the apparent mismatches between lights in both photos. You can see some lights have clearly-observable time-varying behavior. These are LED-based lights. Some of these can be seen to change from one pattern to another. This corresponds to a visual change in brightness (or on-to-off, or off-to-on), as you would expect with emergency lights. The two white flashes at the left come from xenon discharge lamps mounted in the vehicle's tail lights. These flashed just once during this exposure. (1/10 sec)

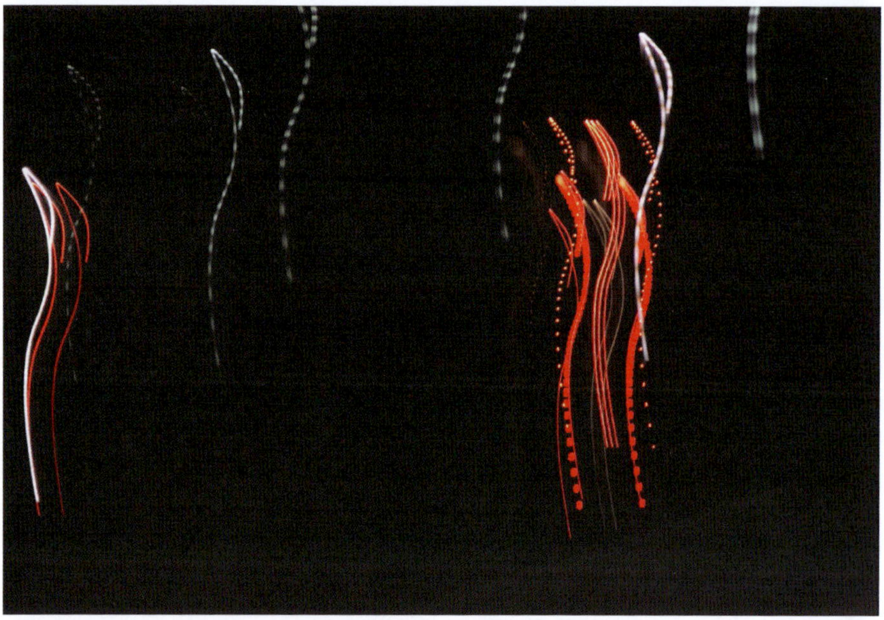

Figure 121: A beautiful swept picture of an ambulance on a street. The ambulance (which is represented by the cluster of red lights on the right) was not running its emergency lights: all the lights here are the red tail lights on the back of the vehicle, including ones that are mounted at the top. Being LED lights, they are strobed rapidly, just like we saw with the Audi cars earlier in this section. See Figure 58 for an alternative treatment of this image. (1/4 sec)

Interior lights

The most common form of interior lighting – at least in 2015 – is the incandescent light-bulb. Sadly (for the reader of this book, at least), it's not very interesting. However, other forms of lighting are much more engaging, especially the ordinary fluorescent tube.

Incandescent lighting

Incandescent lighting – in other words, the humble tungsten light-bulb – is probably the most common type of lighting in the home, though it's beginning to be phased out in favor of compact fluorescent lighting or LED lighting.

Figure 122: The ubiquitous incandescent light bulb, now finally an endangered species.

Incandescent lights are driven by the alternating current (AC) electrical supply, so you may imagine that you would see time-varying behavior. However, it turns out that this is not the case: incandescent bulbs show very limited variation in brightness.

The reason for this is that these bulbs operate by heating a metal filament to very high temperatures (about 2000°C / 3600°F), which is hot enough that it glows white hot. Because the bulb works on both halves of the AC cycle (positive and negative), it receives 120 (100 in Europe) pulses of power per second. This means that the filament does not have time to cool down enough to stop glowing before the next pulse of electricity. The filament's heat capacity keeps it hot and glowing even though the AC power is not constantly present. It's a little like a flywheel which keeps an engine rotating smoothly between the ignitions in the cylinders.

The slight variation in brightness that does exist is usually too subtle to be discerned when vision jiggling; indeed, a non-varying trace is often a sign that you're looking at an incandescent light.

Cameras can sometimes detect the variation; and in fact the dimmer the lamp is, the more likely you will be able to capture a trace that shows variation.

Figure 123: A swept photographic trace of incandescent light leaking around the door of a closet. The subtle horizontal bands show that the intensity of this light bulb is indeed varying periodically at the expected 120Hz frequency (two peaks of intensity in each 60Hz cycle). This particular example was discernible with the naked, jiggled eye, which is unusual for incandescent lighting. (1/10 sec)

A final note: Those of you old enough to remember vinyl record turntables may recall those paper discs you could place on the turntable to help you with adjusting the rotation speed. This is a stroboscopic device which *relies* on the illumination's periodic variation in brightness (100Hz in Europe, 120Hz in the USA). The slight brightening and darkening of the light bulb causes the pattern of black stripes to appear to stand still when the turntable is rotating at the correct speed. If incandescent bulbs showed no time-varying behavior, this device would be useless.

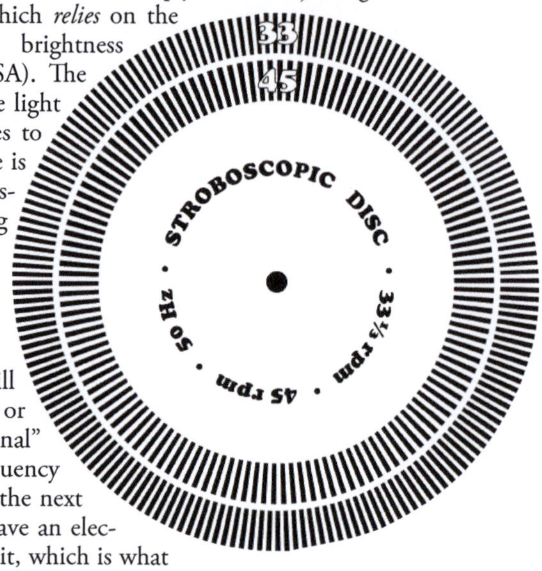

Figure 124: A template for a paper stroboscopic disc for adjusting the rotation speed of a record turntable. This one is designed for incandescent light on a 50Hz AC supply (e.g. in Europe). The "33" band appears to stand still when the turntable is rotating at exactly 33 rpm; the "45" band at 45 rpm.

Traditional fluorescent lighting

Most fluorescent tube lights that you will find in a home (typically in the garage or workshop) are of what I call the "traditional" type, as opposed to modern high-frequency fluorescent lights which are covered in the next section. Traditional fluorescent tubes have an electromagnetic (coil-based) "ballast" circuit, which is what drives the electrical discharges in the tube, and these tubes can be identified because they typically flash irregularly once or twice when first switched on. Also, they often buzz or hum due to the vibration of the ballast components.

There are still many traditional fluorescent lights around, and they are a great subject for the observer of time-varying light sources.

The most striking thing about fluorescent tubes is not that they change in brightness (which they do), but that they also change *color*. Observe one and you will see that it emits a bright, slightly bluish white for about 50% of the time, and a darker yellow light for the remaining 50%.

Figure 125: A typical trace from a traditional fluorescent tube light. (1/10 sec)

This can be observed best if you jiggle your vision (eyeballs or mirror) along an axis that is perpendicular to the length of the tube. See page 19 for further discussion on choosing the direction of jiggle.

Because fluorescent lighting is emitting a dull yellow light for so much of its cycle, photographing objects under such a light can provide surprises, as the next two images illustrate. See page 35 for more details on how time-varying light sources can affect photography.

- Observe a single fluorescent tube from a distance and see the yellow bands in the retinal trace.
- Take something white, such as a white shoe, and watch it while you wave it rapidly under a fluorescent light. You should see the same yellow bands.
- If you have a fluorescent light above a sink, run water onto a tilted frying pan or skillet and look closely at the colors you can see in the turbulent water as it flows. Yellow and white should both be apparent.

Figure 126: A photograph taken in my workshop under fluorescent lighting. Colors are natural and as expected. (1/200 sec)

Figure 127: A photograph taken just a moment later. In this case, the exposure occurred when the fluorescent light was in its dull yellow phase. Even with the camera's white balance set to "auto", the yellow cast is pronounced. Had the white balance been set to a static value that matched the white phase of the lighting, this image would have been even more yellow. (1/200 sec)

What's happening?

Fluorescent tubes operate by sending electrical discharges through a tube filled at low pressure with mercury vapor. However, due to the quantum physics of the mercury atom, the light that's emitted is ultra-violet, and cannot be seen by the human eye. To turn this into a useful lamp, a coating of fluorescent chemicals is applied to the inside of the tube. These absorb the ultra-violet light and re-emit it at lower energies, which are in the visible spectrum. Several chemicals are combined in this coating, each emitting visible light of a different range of colors; the overall combination is perceived as white light.

When the electrical discharge ends as the AC voltage drops towards zero, some of these chemicals continue to glow for a little longer than other ones; in particular, the yellow-emitting fluorescent chemical glows for longer, which is why there is a period (during the time when the mercury vapor is not emitting ultra-violet light, i.e. in between discharge flashes) when yellow light is still being emitted.

Note that because of the way the human eye perceives color, this light could truly be composed of yellow light, or it could consist of red light and green light together. Both cases would appear to be yellow. Hence, for some fluorescent tubes, one can suppose that three phosphors (fluorescent chemicals) might be active: red-emitting, green-emitting and blue-emitting. When all three are active, one would see white light; if the blue phosphor becomes inactive during non-discharge times but the red and green phosphors remain active for an extended decay time, one would see yellow light.

The following diagram should serve to illustrate what's happening during the AC cycle:

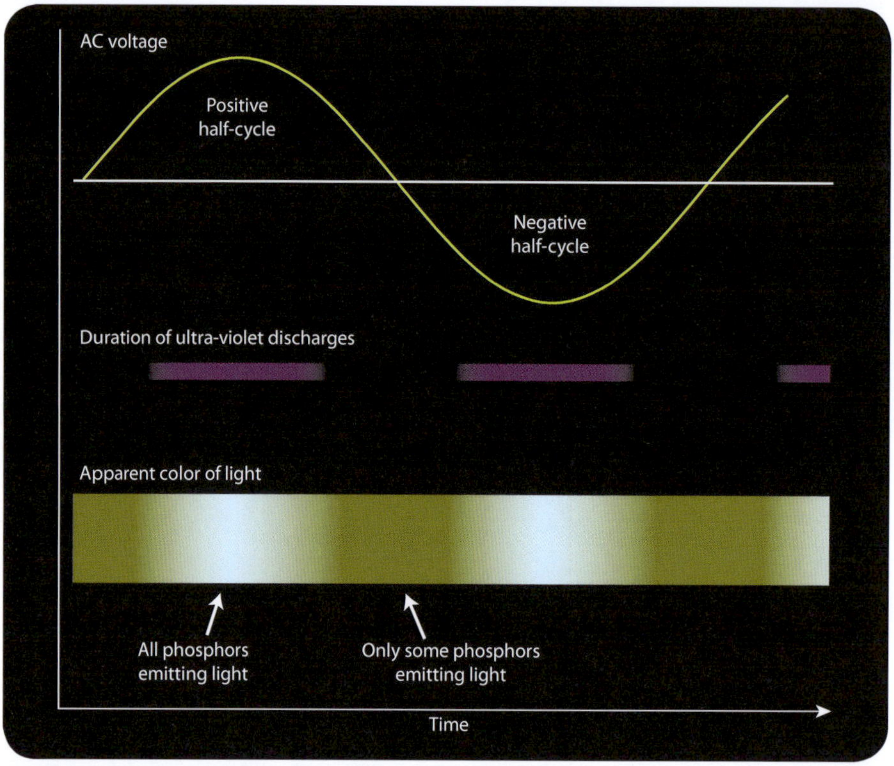

Figure 128: A diagram illustrating the sequence of events in a traditional fluorescent tube. The sine curve at the top represents the voltage of the AC mains supply, for example, 110V at 60Hz (in the US). The purple dashes indicate the periods when the mercury vapor is undergoing electrical discharge and emitting ultra-violet light. The important thing to note is that the discharge only occurs when the voltage becomes sufficiently high (positive or negative). When the AC voltage is closer to zero, no discharge occurs. The white and yellow bar indicates the perceived color of the light from the tube. The white periods occur when all phosphors are emitting; the yellow periods when not all phosphors are emitting. Some phosphors cease emitting as soon as the ultra-violet stimulation ceases; others, which contribute to the yellow color, continue to glow for a short time, essentially filling in the gaps between mercury discharges.

The following images are examples of fluorescent lights and the traces they generate.

Figure 129: A building illuminated by a row of fluorescent tubes.

Figure 130: A photographic trace of the building. The color variations are very apparent, with the tubes showing a quite complex time-varying behavior. Note that the dark yellow or brown bands line up with each other: since the tubes are physically arranged in a straight line, this trace proves that all the tubes are connected to the same phase of the three-phase electrical supply. (1/10 sec)

Figure 131: A building with six fluorescent tube lights along its roof-line.

Figure 132: A truly amazing photographic trace of the lights in the previous photograph. The dark yellow periods amount to *at least* 50% of the cycle, sometimes much more, and additional dark yellow bars and white flashes can be seen. It seems amazing that these lights actually appear white, given the amount of time they spend being *not* white! Of course, the white periods are brighter than the yellow periods (squint at the image if you're not convinced of this, and see how the white bands stand out). Notice that there appears to be some variation in the *phase* of the patterns: from the left, lights 1, 2 and 6 are in phase with each other, but lights 3, 4 and 5 are apparently on a different phase. You may be tempted to think (given what I've written elsewhere in this book) that this indicates that these lights are wired to two phases of the three-phase electrical supply. However, closer inspection shows that the two "phases" are offset by one *half*, not by the tell-tale one *third* of three-phase electricity. For example, the thin yellow line of light 3 is exactly half-way between two thin yellow lines of light 2. This "half-way" relationship reveals that in fact all six lights *are* connected to just one phase of the electrical supply; but that some of the lights are triggering on the positive half-cycle, and some on the negative half-cycle of the AC voltage. In fact, another photograph (not shown here) that I took a few seconds later shows a different pattern, so the triggering is somewhat erratic, as might be expected for these obviously aging fluorescent lamps. (1/10 sec)

Figure 133: This photograph was taken through the front door of a Famous Footwear store at night when the store was closed. We can see two lights in the ceiling. To the naked eye, these lights appear to be identical, both shining with steady white light.

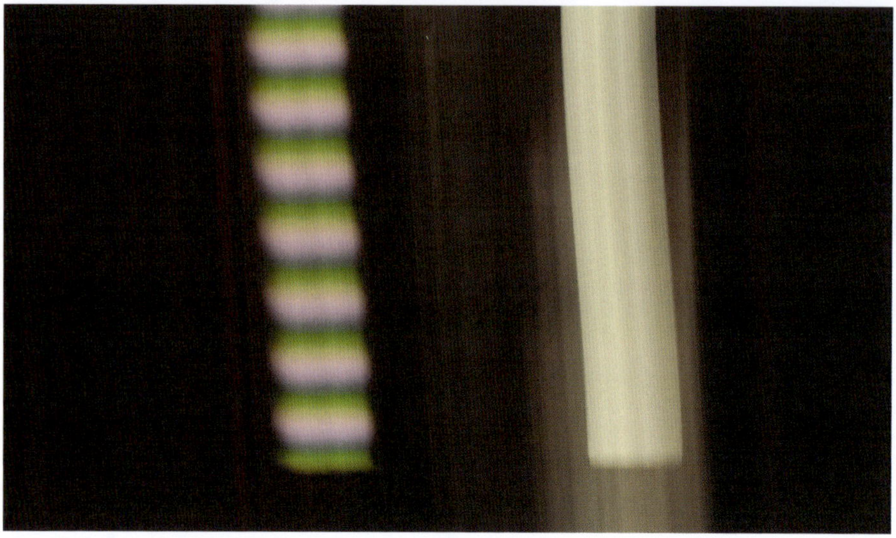

Figure 134: However, the picture changes dramatically when the camera is swept: While the right-hand light is completely non-varying (it may be a high-frequency fluorescent light), the left-hand light is alternating through periods of remarkably saturated pink and green, with some yellow and blue also visible. In fact, it never emits *white* light at all! It *appears* white because if you mix pink light and green light, the eye perceives the result as white, as long as the pink and green keep coming fast enough that the eye can't see the individual periods. And just to be clear: this colored trace is very clearly seen with the jiggled naked eye or a mirror. (1/10 sec)

High-frequency fluorescent lighting

There is a newer type of fluorescent lighting that is common now in offices and stores: high-frequency fluorescent lighting. The electronic circuit which controls the power to the tube operates at over 20kHz (20,000 cycles per second). Such lights start instantly, without the flash or two that ordinary fluorescent tubes usually exhibit when starting; and they are silent (no hum or buzz). They are also more efficient, putting out more light for the electrical power that's consumed.

The only downside of this technological improvement is that this form of lighting offers nothing interesting for the vision jiggler. The light output is constant: it's not time-varying at all. Since the mercury vapor is flashing at 20kHz, the fluorescent materials are pretty much constantly stimulated, and there is no time for the development of the dull yellow phase seen in traditional fluorescent lighting.

Compact fluorescent lighting

Compact fluorescent lights are becoming increasingly prevalent in homes and other buildings due to their relative efficiency compared to incandescent lighting. Indeed, some governments are requiring the phasing-out of incandescent light bulbs over the next few years. From an energy efficiency perspective, this is a good thing; however, the downside is that since these new "bulbs" contain mercury vapor, they do represent a toxic hazard should you drop and break one.

Figure 135: A compact fluorescent lamp.

The only other thing to note is that almost all such lights do not show any observable time-varying behavior, and are not very interesting to the reader of this book. This is due to the use of an electronic ballast circuit, which is the component that drives the electric discharges in the tube. For this type of lighting, electronic ballasts can operate at over 40kHz (40,000 times per second), which means that the fluorescent materials are constantly stimulated, giving no time for the dull yellow phase to develop. 40kHz may also be faster than even the fastest eyeball-jiggler could see, because any variations along the retinal graph would be too close together to distinguish.

Discharge lighting

Figure 136: A mercury vapor discharge light as used in a gymnasium or commercial building.

You are not going to find mercury or sodium discharge lights in your home (or anyone else's); and not in your office, either. This kind of light is found in high-school gymnasiums, basketball courts, factory floors and the like. A typical light of this type is shown in Figure 136. These lights are often mercury vapor lights, and therefore they behave much the same as street lights of the same type, covered on page 55.

Typically, these lights emit a strong greenish-white colored light for most of the time, with a dull pinkish-red light emitted between flashes. It may be possible to observe this by vision- jiggling, except that the strong brightness of the light can overwhelm the eye and make it hard to see the color of the dark part of the cycle.

- Observe the time-varying behavior of these lights by vision-jiggling.

- Next time you are at an indoor sporting event, such as a gymnastics meet, see if you can see light and dark bands on a gymnast who is moving quickly. To achieve this, you will need to allow the gymnast to pass across your vision without letting your eyes follow him or her.

Finally, it's worth pointing out that sensitive cameras can detect the pinkish-red color just fine. See the section on "side effects of time-varying light sources on photography" on page 35 for more details.

Light Emitting Diode (LED) lighting

A new type of illumination that is rapidly becoming prevalent in the market is LED-based lighting. It has many advantages over traditional incandescent lighting, not least its far superior energy efficiency. However, it has taken a few years for mature products to come to market. In its most usual form, it's a bulb with a cluster of white LEDs in it.

Figure 137: A collection of LED lamps for the US market.

As far as time-varying behavior is concerned, there is a wide range, which makes white LEDs an interesting target for observation. However, it is increasingly the case that white LED light bulbs are "DC driven", which means that the LEDs are constantly illuminated, with no variation. This is desirable, as it eliminates any flickering that some people may be sensitive to.

Figure 138: *(left)* A typical white-LED light-bulb replacement designed for interior lighting. This photograph was under-exposed for clarity. In reality, the bulb is bright.

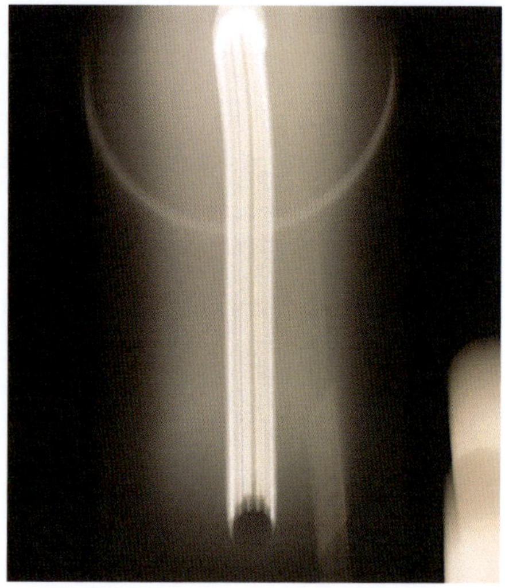

Figure 139: *(right)* A photographic trace of the light in the previous figure. The LEDs in this product are DC driven: there is no variation in brightness. This is increasingly typical for LED products for the home. (1/10 sec)

Other LED lights are time-varying, but what you will see depends on a combination of two factors: (a) the electronic circuit which powers the LEDs (known as the "driver"), and (b) the physical properties of the chemicals in the LEDs themselves.

The driver's job is to convert the AC electrical supply (for example 110V at 60Hz in the USA, 220V at 50Hz in Europe) to a low-voltage supply which powers the LEDs. Some drivers simply reduce the voltage, with the result that the LEDs will flash 60 (or 50) times per second, with the "on" time about the same as the "off" time (this is called a "50% duty cycle"). Christmas lights are often like this.

Other drivers might "invert" or "rectify" half of the AC cycle, with the result that the LEDs will flash 120 (or 100) times per second.

Other drivers might use an electronic circuit to flash the LEDs at a higher frequency, with the duty cycle (the ratio between "on" time and "off" time) set to a specific value in order to control the brightness of the light. In some cases, the "on" period may be very short. And still other drivers may operate a more complex cyclic behavior for reasons of their own.

On top of this range of driver behaviors, the picture is complicated further by the behavior of the phosphors in the LEDs. It's not possible to create an LED that emits white light as such, so white LEDs are implemented by coating blue LEDs with chemicals which absorb some of the blue light and re-emit it as red, yellow and green light through a process called "fluorescence". The resulting mixture of colors appears white to the human eye. It turns out that some of these chemicals can continue to emit light even after the pulse of blue light from the LED has finished. This is called "phosphorescence". In white LEDs, the time taken for phosphorescence to die away (the "decay time") can vary from zero to a few milliseconds.

LEDs for which the decay time is close to zero will have clearly-defined ends to the pulses of white light, and obvious dark periods. LEDs for which the decay time is longer will have a strange, tapered "tail" of white light. These are particularly interesting to observe. This tail can occupy most or all of what would have been the dark period had there been no phosphorescent decay.

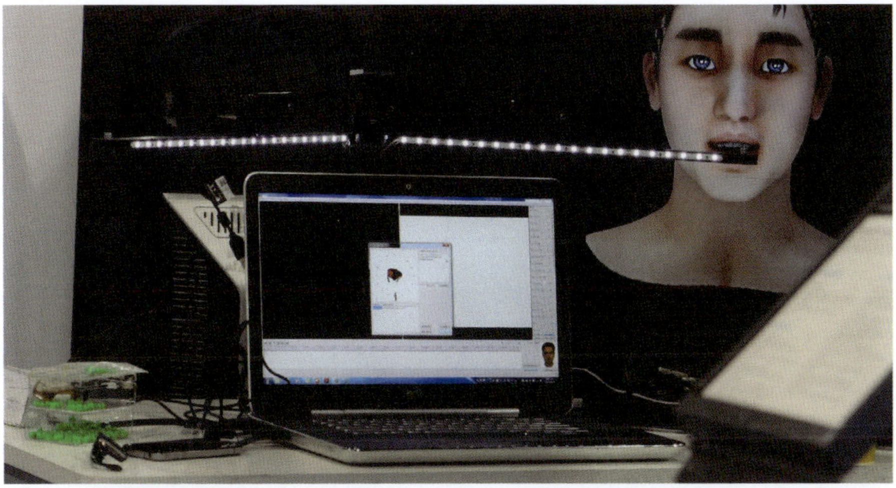

Figure 140: A photograph of a demonstration at a technology trade show. I would like to draw your attention to the two bars of white LED lighting above the laptop computer. In this case, these lights are used to illuminate the face of whoever stands in front of the camera in the center.

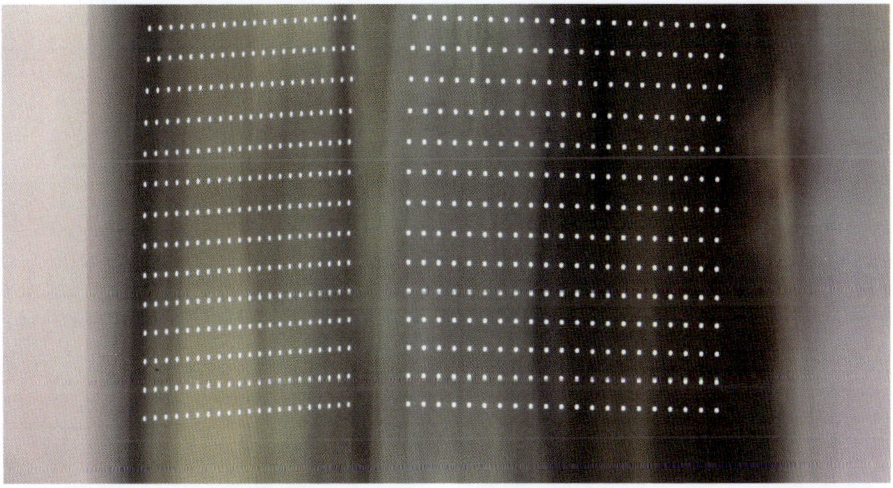

Figure 141: A trace of the scene shows the obvious time-varying behavior of the LEDs. The electronic circuit (the "driver") which powers the LEDs delivers just a short pulse followed by a long dark period. Since there are no visual tails, we can infer that the phosphors have a decay time of close to zero milliseconds. This type of time-varying behavior is very easily seen with vision jiggling. Strips of LEDs like this are often used as under-shelf lighting in stores. (1/10 sec)

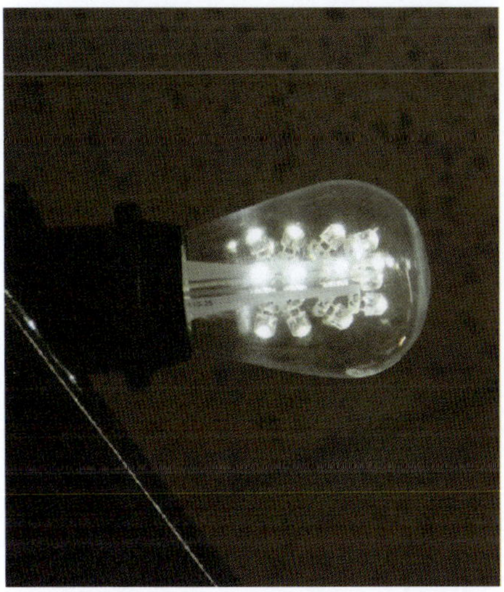

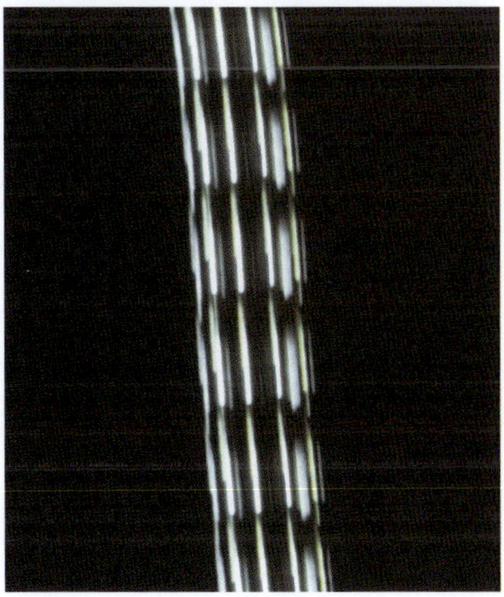

Figure 142: *(left)* An LED lightbulb consisting of a "tree" with many white LEDs. This photograph was under-exposed for clarity.

Figure 143: *(right)* Close inspection of the trace from this bulb shows "tails" of white light, which appear to taper. These tails are due to the relatively long decay time of the phosphors. The tapering is a little hard to explain, but it may simply be that as the light output decays, visible emission converges to a smaller point. (1/10 sec)

- For any given white LED light, determine if the output is time-varying or not.
- For time-varying white LEDs, observe whether there's a phosphorescence decay tail or not.
- Walk into a lighting store and see if you can identify any lights that are time-varying. Most will not be.

LED accent or decorative lighting

The versatility of LED lighting has resulted in a wide range of products which fulfill the function of accent or decorative lighting. In many cases, these lights can be set to whatever color is wanted, or can be set to cycle through a sequence of colors. To achieve this feature, a combination of red, green and blue LEDs are used, and the relative intensities are modulated to achieve the desired perceived color. For example, mixing red and green light gives yellow; reduce the green intensity and you will see orange. As we have seen elsewhere in this book, this type of modulation is achieved by flashing the LEDs on and off rapidly, with the on-duration to off-duration ratio setting the apparent intensity. Naturally, this is a good subject for the observer.

Figure 144: A night-time photograph of the window of a gift store. We are interested in the pink decorative LED rope that's along the edges of the window. It changes color all the time.

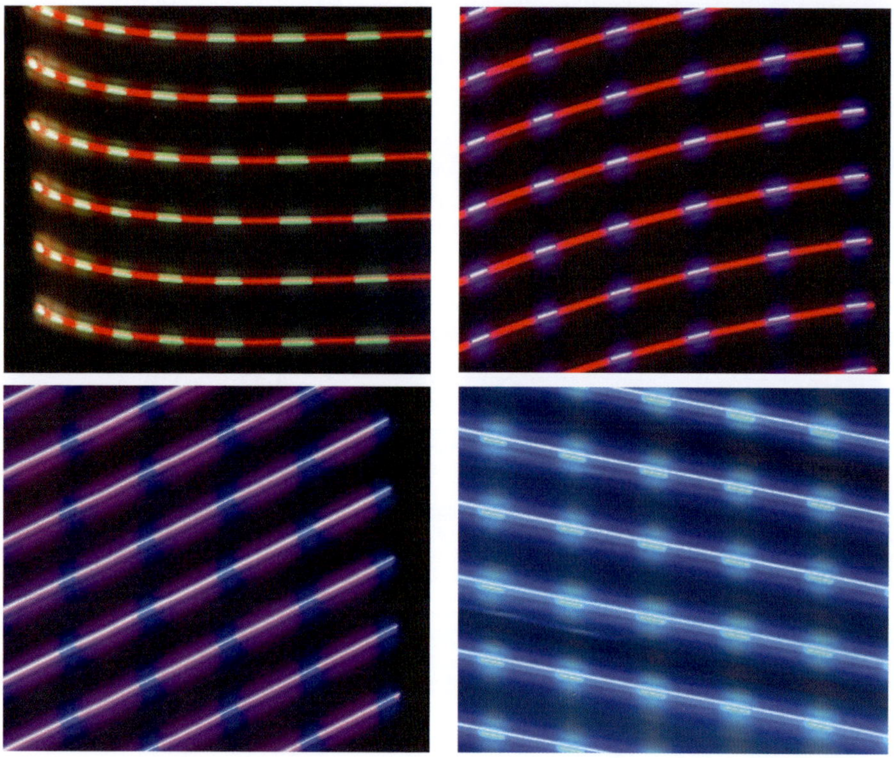

Figure 145: Four photographic traces of the LED rope in the previous figure. The camera was zoomed in close so that we can see the individual LED clusters extended into colored lines. It's obvious that the colors and patterns change as the overall color of the rope changes. Most lights of this type use red, green and blue LEDs. Here, it looks like there may also be violet LEDs. However, the violet sections are due to the red and blue LEDs being on simultaneously. All images 1/10 sec

Electronics

It would be interesting to count how many electronic devices are in a typical household these days, from TVs to cameras, toys to computers; not to forget hand-held devices and game consoles. Most of these emit light in one form or another; and a lot of that light is time-varying. Because different types of displays and indicators can be found on different devices, this section is mostly organized by display type, such as LED indicators or numeric displays. Devices which are treated individually are TVs, computer monitors and projectors.

Numeric displays

Numeric displays are everywhere on consumer electronic devices; on microwave ovens, CD, DVD and video tape players, alarm clocks, ovens and refrigerators, to name the most obvious few. Of course, some of these displays are not purely numeric as such: they may also show letters and symbols showing the status of the device. They appear in many colors; most are green or blue, but red is also common; and I have even seen an alarm clock with a different color for each digit (pictured here).

These displays come in three main varieties: LCD (liquid crystal display), LED (light emitting diode) and vacuum fluorescent.

We will ignore LCD numeric displays in this book, as their time-varying behavior – if any – is hard to observe in most cases, although the back-light may have some interesting characteristics.

Figure 146: An alarm clock with a different color for each digit.

Vacuum fluorescent displays are almost always blue or green; LED displays can be almost any color. Both kinds of displays are almost always time-varying, and are interesting and easy to observe.

The most interesting feature of these displays is that they usually light up only a sub-set of the segments at any time, for example one digit at a time. This is probably done to save on the cost of the circuitry that drives the display. Since the digits are operated by strobing (a short flash of illumination followed by a relatively long period of darkness), it's easy to light up each digit in turn and still give the appearance of being constantly lit.

Jiggling your vision at a typical numeric display will produce a retinal trace like the one shown in Figure 148. Note that the digits form a staggered pattern. From this pattern it is possible to deduce the order in which the digits are illuminated. In the case of Figure 148, it is right-to-left.

Some displays break down the timing at an even lower level than by digits: they flash individual segments on and off at different times (see Figure 149).

For any given numeric display, see if you can determine the order in which the digits are illuminated. Some are illuminated in order, left to right or right to left; others are staggered. For extra credit, if you find one that's left to right or right to left, try to identify which is the case. This takes a little thought, and is achieved by being aware of when you are jiggling upwards or downwards, and what you are seeing as you jiggle in each direction. Single sweep jiggling may help (see page 21).

Look out for some clocks which have DC-driven displays. These are always LED displays (never vacuum fluorescent), usually red, and are designed to operate in bright environments. These have no time-varying behavior because the LEDs are illuminated all the time.

Figure 147: An LED display on a domestic oven.

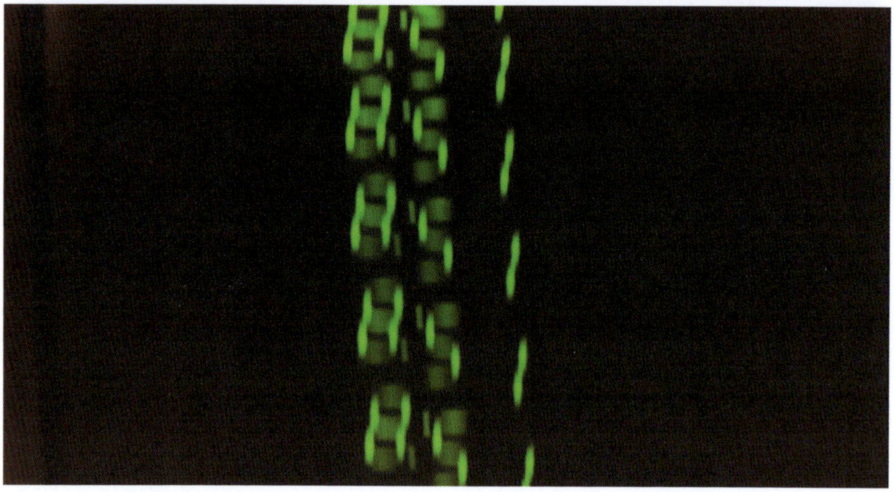

Figure 148: A photographic trace. Since the camera was swept downwards, it's possible to determine that the digits were illuminated in a right-to-left sequence. Note, however, that the digits do overlap (if you draw a horizontal line through them, say). This indicates that in this case the digits were not illuminated one at a time. (1/10 sec)

And what about that multi-colored alarm clock? Its trace looks like this:

Figure 149: A photographic trace of the alarm clock shown in Figure 146. Since the camera was swept downwards, lights which line up horizontally in the trace are on the whole illuminated simultaneously (I say "on the whole" because the digits do also have height to them, which complicates the trace a little - see page 18). We can say with some certainty that the red dot on the left flashes at the same time as the upper left segment of the green digit, both blue dots of the colon (which are almost smeared together), and the right-hand segments of the red digit. The other segments are illuminated a moment later, when the first set has gone dark. It broadly shows an alternating pattern, with just two "phases" to the illumination. (1/10 sec)

CRT televisions and monitors

Cathode Ray Tube (CRT) televisions and computer monitors were common for many years, and have only relatively recently been supplanted by plasma and LCD flat panel displays. You can tell which televisions and monitors use CRTs: the front of the screen is not flat, and the device is usually quite deep front-to-back, and heavy. They also usually take some time to warm up.

By contrast, plasma and LCD displays always have a completely flat screen (except for the new curved screens), and are usually shallow front-to-back. These are covered in the next sections.

Figure 150: A typical CRT television set, now a dying breed, though still common in motels.

Figure 151: A (rather old) CRT computer monitor, showing an image that you will see in later figures in this section. This photograph was taken at 1/80 second, and the screen is refreshed by the computer 75 times per second (75 Hz). This means that an eightieth of a second is a little too short a time to capture the whole "frame", as it's called. This explains the dark band that runs across the monitor near the top of the image. In fact, artifacts like this are a common side-effect of the time-varying behavior of CRT displays on photography and videography - see page 35.

CRT displays are among the most interesting light sources for the vision jiggler because they have complex time-varying behaviors which lead to striking traces. With careful observation you can see how they are operating, if you know what to look for and how to interpret what you see.

CRT televisions make the best targets, because they operate relatively slowly and are easiest to observe. CRT computer monitors tend to operate at higher frequencies, and are less rewarding. Of course, CRT televisions are becoming a rarity in countries like the USA; but if you can find one, you can expect some very interesting observations. Many motel rooms still have CRT televisions at the time of writing. This section gives you a(nother) way to have fun in a motel room with the lights off!

Cathode Ray Tubes operate by directing a beam of electrons (called a "cathode ray" because the cathode is the emitter of electrons in the tube) at a glass screen that has phosphorescent chemicals deposited on the back surface of the glass. Phosphorescent chemicals emit light when irradiated with electrons. There is a vacuum in the tube; without it the electrons would not reach the screen. This vacuum is one reason that the screen is always curved to some degree – a flat screen would be at greater risk of imploding.

Electrons that are moving can be deflected using magnetic fields, and so CRT displays are equipped with electromagnetic coils which deflect the beam of electrons in such a way as to systematically traverse the screen, effectively "painting" an image in light on the screen. The intensity of the beam is modulated too, to achieve lighter and darker parts of the image; this modulation is synchronized with the systematic traversal of the screen, and the image is built up in this way.

Actually, color TVs and computer monitors have three beams of electrons, and a clever system of masking allows one beam to stimulate phosphors that emit red light, another beam stimulates green-emitting phosphors, and the third beam stimulates blue-emitting phosphors. While interesting from a color science perspective (which is another book in itself), this three-beam implementation doesn't change anything as far as time-varying observations are concerned, so for this discussion we'll pretend that there is just one beam of electrons (as would in fact be true for black and white TVs and monochrome monitors).

In order to compose an image, the path traced by the beam follows a pattern of horizontal lines. Each line is traced from left to right, and then the next line down the screen is traced. There will be several hundred lines. For example, a standard analog television set in the USA has 480 lines; in the UK it has 625 lines. A computer monitor may have 768, 800, 1024, 1280 or some other number of lines.

What's impressive is that the *entire screen* is traversed in this way for every frame of the TV transmission – and (depending on your country) there are 25 or 30 such frames *per second!* (Strictly speaking, an analog TV broadcast is interlaced, which halves the number of lines in each actual frame, but this doesn't change anything for our purposes.) This means that the electron beam is tracing out all those lines, covering the whole screen from top to bottom in just a 25th (or a 30th) of a second. And a computer CRT monitor can refresh the screen anything from 60 to 100 frames per second (usually 60 or 75).

Clearly, we have some interesting time-varying behavior here, and the seasoned vision jiggler will be able to unpack it and see it in action.

What's really interesting is that although the image on a TV screen appears to be constantly present, this is an illusion caused by persistence of vision: the retina is stimulated by the light it receives from the screen, and is still responding when the *next* frame is painted by the electron beam. This is a useful behavior of the eye, because otherwise TV broadcasts would require many more frames per second in order not to appear to flicker. Like I said in my introduction, persistence of vision is your friend (page 11).

In actual fact, though – and your observations will prove this – *most* of the TV's screen is actually dark at any moment in time! The only bright part is the line that is being traced right now, and probably a few of the lines above it that were traced very recently. If a camera's shutter speed is faster than the time it takes to paint the whole screen then the camera will capture only part of the painted image and the rest of the screen will appear dark. Figure 162 and following show exactly this effect.

If you're astute, you will have realized by now that in order for the TV image to be built up correctly on your retina, it's a *requirement* that you keep your retina mostly still. That way, you get to see the sequence of images that make up the TV transmission. However, if you're interested in seeing the TV set's time-varying behavior, you will need to move your retina around (eyeball jiggling) or cause the image of the TV to move around on your retina (mirror jiggling).

- Watch a CRT television from the far end of a darkened room. Sweep your vision sideways across the TV, first as a single sweep to the left, then as a single sweep to the right. Do this by looking at a point that is off to one side of the TV, and then rapidly and smoothly shifting your gaze to a point off to the other side of the TV.

You will see:

- Multiple images of the screen; this shows that the screen is indeed being repainted frequently.
- These images are slanted! When you sweep your vision to the left, the images slope towards the left, and when you sweep to the right, the images slope towards the right.

These photographs illustrate what you will see:

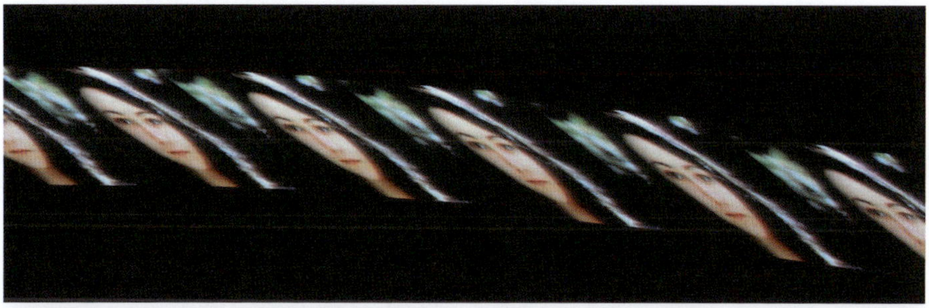

Figure 152: A left-swept trace of the monitor shown in Figure 151, captured in a dark room. (1/10 sec)

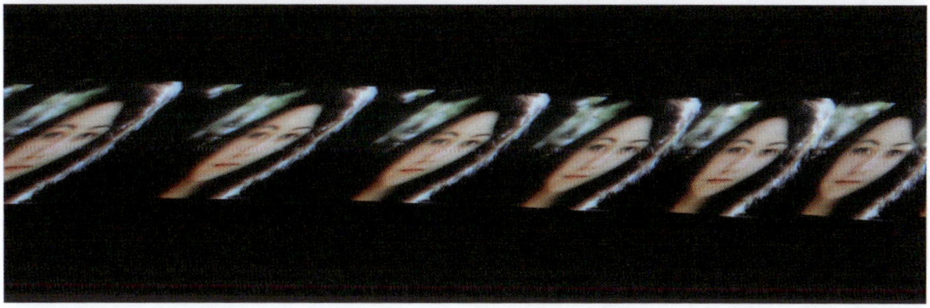

Figure 153: A right-swept trace of the same. (1/10 sec)

Why is this happening?

Remember that the image is being "painted" from the top down, line by line. This means that if your eyes are sweeping to, say, the left, the location of the TV's screen is sweeping across your retina. The top line forms an image on your retina, and then it's time for the next line, but your eyes have moved sideways a little (due to you sweeping your vision), so when the next line forms an image it's not exactly underneath where the first line was imaged; and so on. The entire stack of lines gets skewed as each line is imaged in a slightly offset position from the line above.

✍ It follows that the faster you sweep your vision, the more tilted will be the images of the TV. Try it.

This effect is visible because it takes an appreciable time (a 25th or a 30th of a second, for TV, a 60th or a 75th for a computer monitor) for the entire screen to be painted, and your gaze has moved quite a long way in that time.

✍ Now try some downward vertical sweeps. Do this by looking at a point above the TV and then smoothly and rapidly shifting your gaze to a point somewhere below the TV.

You will see a range of effects depending on how fast your sweep is:

- If your sweep is at a certain critical speed you will see a bright band of white light. This is because you essentially watched as the TV painted horizontal lines down its screen; your eyes were moving at just the right speed to follow the lines as they were drawn! This appears as a bright band because all the lines on the TV's screen were imaged on the same place on your retina. Figure 156 illustrates this case.

- If your sweep is a little slow, you will see the image on the TV somewhat compressed vertically. This is because your vision is *somewhat* watching the progress of the lines as they get painted down the screen, but they're running ahead of the motion of your retina. This results in the lines being painted on your retina in the right order, but spaced closer together than they would be if your eye was not moving. Figure 154 is an example of this case.
- If your sweep is really fast, you will see the TV's image upside-down! This is because your vision has run *ahead* of the lines as they're getting painted, and your retina has moved on faster than the lines are getting imaged, so they end up in the wrong order! See Figure 160.

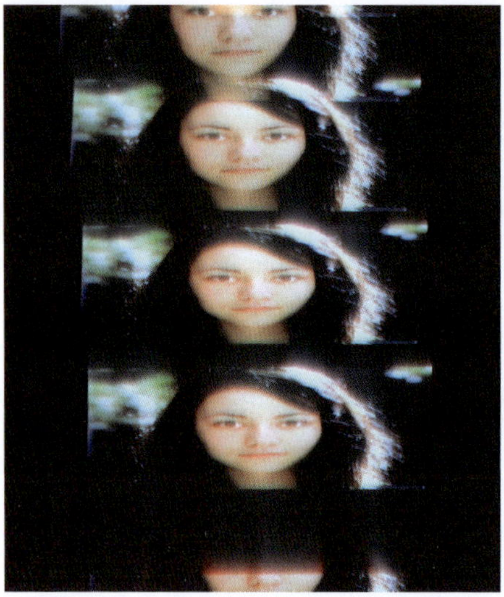

Figure 154: *(left)* Here we have a sequence of photographic traces, all taken with a single downward sweep of the camera, with an exposure time of 1/10 second. These pictures show the effect of faster and faster sweeps. This one is relatively slow; the image looks just a little compressed vertically, and the separation between the images is very small. To achieve the fast sweep speeds, I stood at some substantial distance from the monitor (maybe 15 or 20 feet) and zoomed in (75mm with a 1.6x field of view crop factor on the Canon EOS 40D).

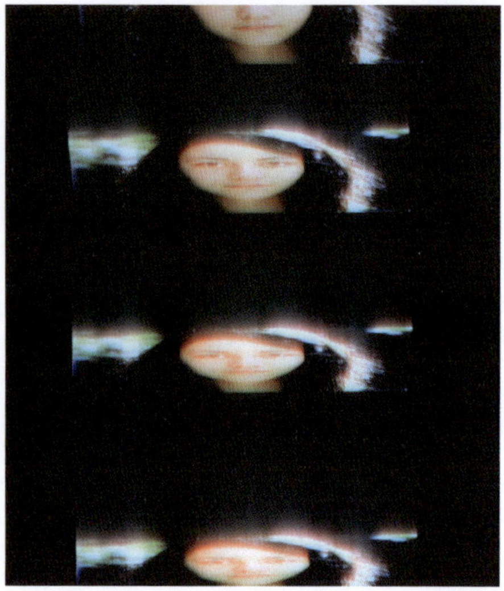

Figure 155: *(right)* A slightly faster downward sweep of the camera. The images are more compressed vertically, and there is more space between them, due to the increased speed of the camera's downward sweep: it travels further before the next time that the monitor's screen is lit. (1/10 sec)

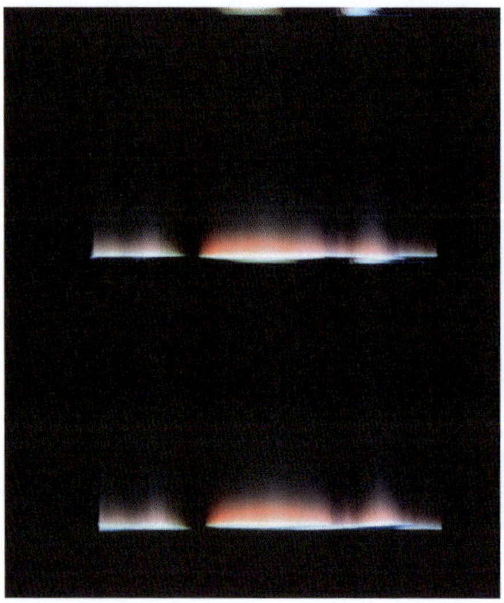

Figure 156: *(left)* A faster sweep. In this case, the camera's view is descending at the same rate that the illuminated line is descending the monitor's screen, so we essentially see a complete collapse of the picture onto one line. The glow that's above the line is due to the fact that the phosphors take a few milliseconds to stop emitting light after the electron beam has stopped exciting them. The separation between these images is greater than before due to the faster sweep. (1/10 sec)

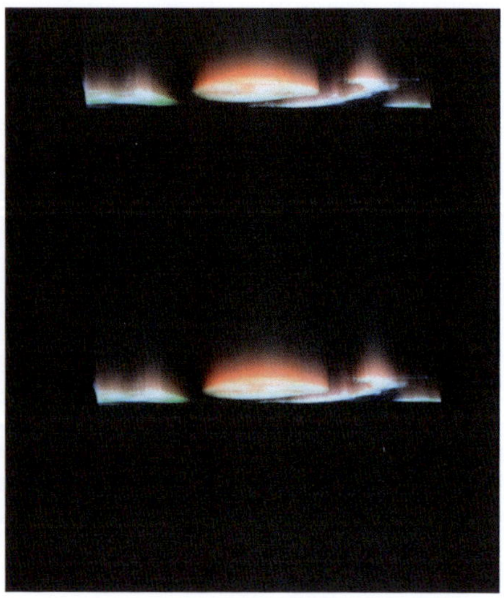

Figure 157: *(right)* Sweeping even faster: the bright lines are now expanding again, but if you look closely you can see that the very compressed images are *upside down!* This is expected, and can also be seen by vision jiggling; see the text. (1/10 sec)

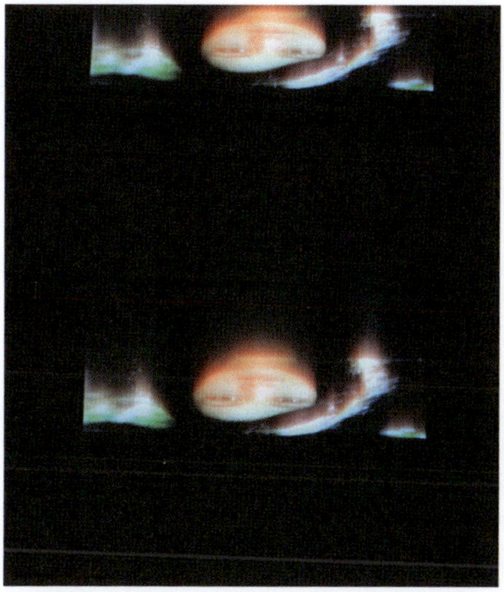

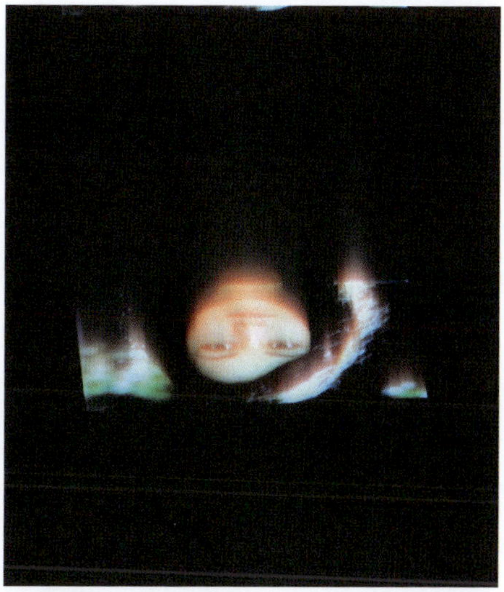

Figure 158: *(left)* Faster still... The upside-down images are less compressed vertically; and the space between them continues to grow as we sweep faster. (1/10 sec)

Figure 159: *(right)* Even faster: Less compressed vertically, and the separation is now large enough that we see only one image. In other words, the camera's view swept past the monitor so fast that the monitor only had time to paint the screen once during the sweep, which it does (for this monitor) every 75th of a second. I was moving the camera pretty quickly by this point! (1/10 sec)

Figure 160: *(left)* Finally, with the camera sweeping about as fast as I could make it go, we have a complete upside down image. This means that the camera's view was sweeping downwards at about twice the rate that the image was being painted line-by-line on the screen. (1/10 sec)

Figure 161: *(right)* An upward sweep causes the image to be *expanded* vertically. Note that there's a second image overlapping the top of the main image. (1/10 sec)

Figure 162: *(left)* To further illustrate the transience of an image on a CRT TV or monitor, I am including this series of images. The camera was not swept, but the exposure time (the time that the camera's shutter was open) is decreased from photo to photo. This image was shot at 1/100 second. Since the monitor takes 1/75 second to paint the full screen, the camera's exposure was too short. This is why part of the screen is dark.

Figure 163: *(right)* This photo was taken at 1/320 second.

Figure 164: *(left)* This photo was taken at 1/800 second.

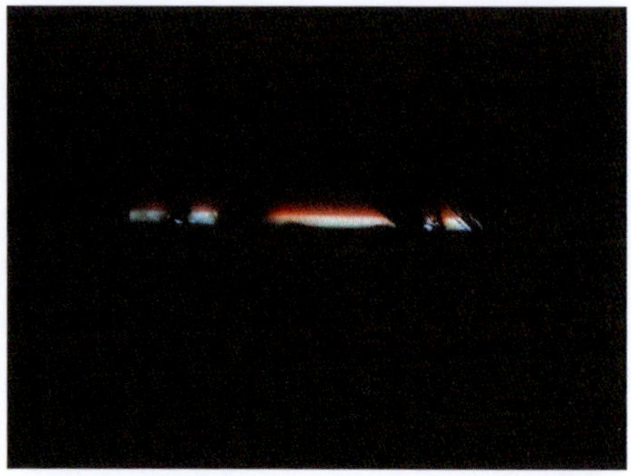

Figure 165: *(right)* This photo was taken at 1/2000 second. During this time, very little of the screen was illuminated by the electron beam painting the phosphors.

Now repeat the downward sweeps with the lights turned on in the room. What you are looking for is a little subtle to see, but keep trying and you'll see it: you will see the TV with a dark tube with a bright band across it. The TV will almost look as if it's turned off – apart from the bright band, of course. This is proof that the TV's screen is mostly dark, and that only a small part of it is actually emitting light at any one time. What's more, this is a consequence of the sophistication of the human vision system: it will try to make sense of what it sees, and so the TV itself will appear mostly unblurred: see page 27.

Now turn the lights off again for one last observation: upward sweeps.

Remember how a downward sweep could give the impression of squashing the TV's image vertically? An upward sweep has the opposite effect: it stretches it. In fact, if you sweep fast enough, you will see the individual lines that make up the image that's on the screen! (This is true for a TV, which has only 480 or 625 lines, and a fairly slow frame rate (30 or 25 Hz); for a monitor with, say, 1024 lines and a refresh rate of 60 or 75 Hz, you may not see the lines. This is the case with the photograph in Figure 161.)

This is happening because of the way your retina is moving while the lines are being generated on the TV screen. By moving your view upwards, each line is a little "too far down" relative to the previously imaged line, and so the lines are separated from each other and the image appears stretched.

Plasma televisions

Figure 166: A photographic trace of a white bar (see Figure 168) displayed on a plasma TV. Although the bar appears steadily white to the naked eye, it is in fact cycling at 60 Hz through a range of whites, yellows and blues, exhibiting complex variation over time. (1/10 sec)

Plasma TVs are now also outdated, having been replaced by LCD televisions (see the next section). In fact, as I write this, I recall hearing a recent news article that plasma TV production will finish soon.

To some extent, plasma TVs are a little like fluorescent tubes. However, every illuminated cell on the screen (a "sub-pixel") is like a tiny fluorescent lamp in its own right, using a gas discharge lamp to generate ultraviolet light and phosphor chemicals to convert that light to red, green and blue.

It turns out that the time-varying behavior of a plasma TV is remarkably complex, as Figure 166 illustrates. I have not researched the reason for this complexity at this time.

LCD televisions and monitors

Televisions and computer monitors based on LCD (Liquid Crystal Display) technology are not quite as interesting for the vision jiggler as their older CRT counterparts, though some surprising examples can be found. "Liquid crystal" refers to the array of tiny cells in the glass which contain liquid crystal material, which, under the control of the electronic circuits in these devices control how much light gets through. The light itself comes from a "backlight", which is often a collection of fluorescent tubes in the back of the TV or monitor. Newer products use a different type of backlight: light emitting diodes (LEDs). For the observer of time-varying light sources, the liquid crystal cells (which comprise the pixels) are not interesting; however, the backlight can be worth observing.

We have seen elsewhere in this book (page 70) that fluorescent tube lights can be quite interesting for the observer of time-varying light sources; and to an extent, that's true of the fluorescent backlights in LCD TVs and monitors. However, they tend to be much better behaved, usually with only a subtle variation in brightness. There are exceptions: Hewlett-Packard's 30-inch display (model ZR30w) is quite startling (Figure 171). At full brightness, it shows a constant white with no discernible variation in brightness. But when the brightness is turned down, the backlight cycles through pink, orange, dark blue and green, with no *white* at all! Thanks to the behavior of the retina, this cycle of colors combines to appear white – but is visible when vision jiggling. A good technique for observing the time-varying behavior of backlights is given later in this section.

Figure 167: The HP ZR30w 30 inch LCD display. I have one of these on my desk.

TVs and monitors with LED backlights are becoming more common now that the technology for white LEDs has matured. These products can be more interesting to jiggle at than fluorescent tubes (with the exception discussed above). The reason is that the brightness of LEDs is adjusted not by turning down the current as you might expect, but by switching them on and off rapidly. If they are on more than they are off, the perceived brightness is higher than if they are off more than they are on.

In some (very rare) cases, the backlight's LEDs are not white, but are a collection of red LEDs, green LEDs and blue LEDs. (This is done to enable extreme color accuracy in certain monitors.) Products in this category fall into two classes:

- the red, green and blue LEDs are pulsed simultaneously;
- or the red, green and blue LEDs are pulsed sequentially, but rapidly so that the eye doesn't (usually) detect that the backlight's color is cycling all the time. There may be overlaps, which would give rise to other colors such as yellow.

The vision-jiggler can see all these effects very easily.

To make observations of time-varying effects on LCD computer monitors, it's best to create an image which accentuates the effect. My favorite image for this purpose consists of a black background with a narrow white bar across it, like this:

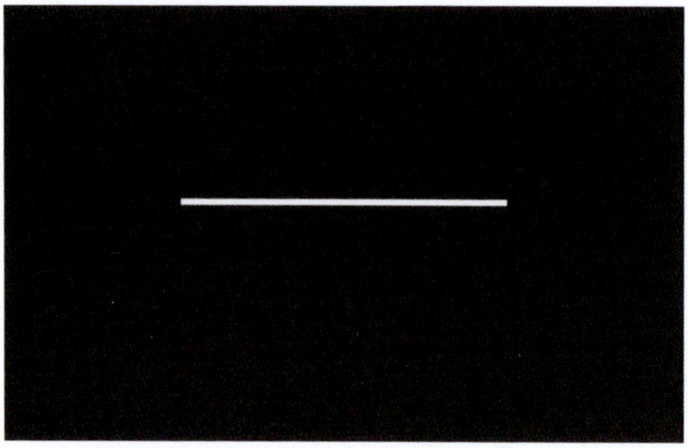

Figure 168: A picture which, when filling the screen of an LCD display, makes it easy to observe the time-varying behavior of the backlight. The traces in this section were taken using a picture similar to this one. If you prefer to jiggle horizontally, use a vertical bar.

Create this image using, for example, Microsoft Paint or Adobe Photoshop; or you can download it from the "resources" section of the web-site: timevaryinglights.com.

Put this test image on the screen of the computer, ideally making it fill the entire screen. (This is easy with a viewer such as IrfanView (www.irfanview.com – just press the Enter key to make the image full-screen; or if you are using Microsoft Paint press the F11 key). Turn out the lights and jiggle your vision up and down as you look at the monitor. The horizontal bar will paint a broad swath on your retina; the time-varying behavior of the backlight will appear as horizontal bands of brightness or color.

Figure 169: A photographic trace of a display which uses white LEDs in the backlight (HP ZR2440w). This trace was created by filling the display with a picture like the one shown in the previous figure. Adjusting the brightness of the display may have an effect on the trace. (1/10 sec)

Figure 170: *(left)* A trace of the HP ZR30w display with the display's brightness set to maximum. This display uses fluorescent tube backlighting, and yet at full brightness no variations are discernible. (1/10 sec)

Figure 171: *(right)* Another trace of the HP ZR30w display, this time with the display's brightness reduced. The fluorescent backlight is showing a striking sequence of colors. Pink and green are dominant, but orange and blue are also present. This image is reminiscent of the one in Figure 134. The amazing thing to realize is that when looking at a white document or spreadsheet on this display it is in fact cycling through all these colors and is never actually white! The retina adds the colors together and you perceive it as white. (1/10 sec)

- LCD displays that have fluorescent backlights will probably show a mild, soft modulation caused by the AC power that drives the fluorescent tubes (the exception being as shown in Figure 171).

- LCD displays that have white LEDs will probably show a more marked modulation, caused by the above-mentioned method for controlling the brightness. Observe what happens when you turn the brightness of the monitor or TV down.

- LCD displays that have red, green and blue LEDs that are modulated simultaneously will probably look very similar to the white LED case, because the eye perceives red light plus green light plus blue light as white; but see if you can observe any colored bands in the retinal trace.

- LCD displays that have red, green and blue LEDs that are modulated in sequence are the most interesting kind to observe. You will see bands of color in the retinal trace. In fact, if you know your color science (additive color, red, green and blue primaries), then by carefully observing the colors in the bands and the order of the bands in the trace, you can figure out the exact sequence in which the LEDs are pulsed, including any overlap periods when, for example, the green LEDs turn on before the red LEDs are turned off. This may be hard to do. An example of a product with this type of display is HP's EliteBook 8770w notebook computer, in the edition with the HP DreamColor display.

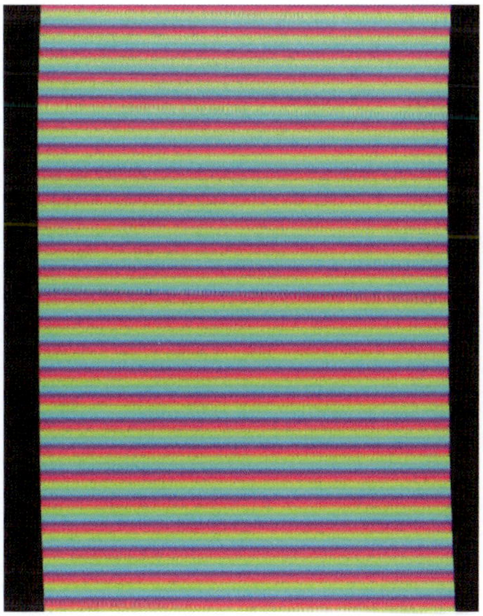

Figure 172: A trace taken from an HP 8770w notebook with the optional DreamColor display. The backlight is comprised of red, green and blue LEDs, which together deliver white light. However, the LEDs are not illuminated at the same time, which means that the illumination cycles through a sequence of strong colors. Again, it's worth pausing for a moment and considering that when you look at a white document on this display it is in fact cycling through all these colors and is never instantaneously white. However, because the eye integrates the colors over time, you see white. In fact, the DreamColor display is designed to deliver very accurate colors for video editors and photographers. (1/10 sec)

Power LEDs

Many pieces of electronic equipment are equipped with a light emitting diode (LED) which serves as a power-on indicator. For the purpose of this book, these fall into four categories: DC, AC, strobed and visibly pulsating. All four types are covered in this section.

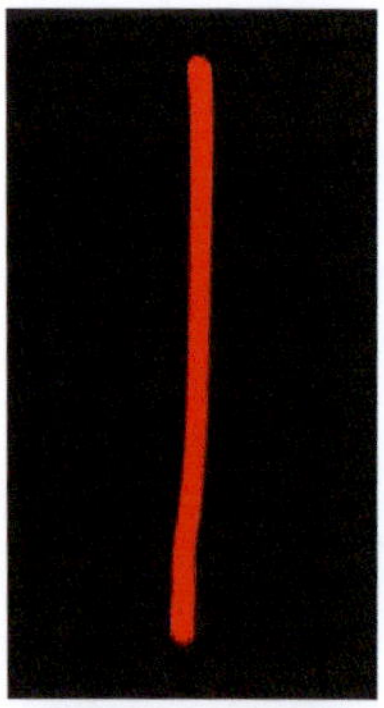

Figure 173: A trace of a DC-driven LED. There are no variations in the trace because this light source is not time-varying.

DC power LED

DC means "direct current", where the electricity that's driving the LED is constant. Naturally, these LEDs are not time-varying, and will produce an unbroken retinal trace.

AC power LED

AC means "alternating current", and is the type of electricity that's delivered through the outlets in your home. Some devices, such as GFCI (Ground Fault Circuit Interrupter) power outlets or electric razors when charging, have a power LED to show that they're activated.

Figure 175: *(left)* A trace of the LED in the GFCI outlet. The dashes have characteristic soft ends due to the sinusoidal rise and fall of the AC voltage. (1/10 sec)

Figure 174: *(right)* A GFCI outlet with a green LED which indicates that the circuit is active.

What's interesting about these LEDs is that they are driven by the AC electricity itself (through a resistor to limit the current). The result is that the LED is time varying in a particular way: the changes from dark to light and from light to dark are relatively gradual. This means that the ends of the dashes in the trace look quite soft.

Notice how "soft" the ends of the dashes are. This is your clue that the LED is powered directly from the AC electrical supply and not from a digital controller circuit.

Also, note that the dark gap between the dashes of light is about as long as or slightly longer than the dashes. In other words, the LED is dark for over 50% of the time. This is a result of the fact that the LED is behaving as a diode, which allows electricity to pass in one direction only. The AC electrical supply goes alternately in one direction and then the other. The LED permits electricity to flow only in one direction, so it only lights up for half the cycle. The other half is dark.

Strobed power LED

By contrast, some power LEDs are strobed, which means that they are switched on and off very fast by an electronic circuit. The reason for this design is that it's difficult to control the brightness of an LED by changing the voltage or current that's driving it, so the solution is to switch it on and off very rapidly; and by controlling the ratio of the "on" time to the "off" time, the *apparent* brightness of the LED can be controlled. This is called modulation.

The result is that many electronic devices have a power LED that flashes very fast. Normally, you can't see it flashing; but vision-jiggling will reveal it to you.

Apple's Macintosh computers often have a white power LED that falls into this category. See Figure 176 for a photographic trace.

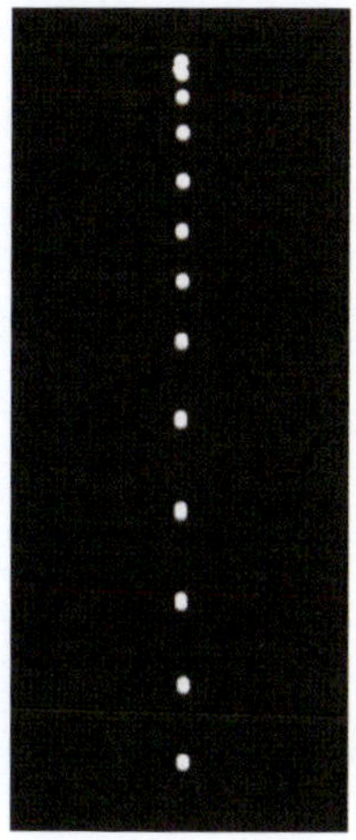

Figure 176: A trace of the small, white power indicator LED on an Apple G5 Macintosh computer. (1/4 sec)

☞ Notice how the bright dashes in the trace have clearly-defined ends. This is because the circuit switches the LED on and off instantaneously, and is very different to an AC-powered LED.

☞ Notice too that the ends of the dashes are typically rounded. Specifically, they are the shape of the LED itself. A square LED will have square (or pointed) ends to the dashes in the trace (according to the direction of jiggle).

☞ By judging the ratio between light and dark periods, you can judge how bright the LED is *perceived* to be relative to its intrinsic brightness if it were driven by a DC current. You may be surprised at just how much of the time the LED is off for.

Visibly pulsing power LED

Some electronic devices have a power LED which slowly pulsates, by which I mean it appears to grow smoothly brighter and dimmer over the space of a few seconds. It's supposed to make the device look "alive". It's a little gimmicky, but quite attractive.

From what you have learned about strobing LEDs, you can probably deduce that the variation in brightness is achieved by varying the ratio of "on" time to "off" time when strobing the LED. For the bright part of the cycle, the "on" time is high and the "off" time low; then the controlling circuit gradually reduces the "on" time and increases the "off" time, and the LED appears to become dimmer.

Many Apple Macintosh computers – those with a white power LED – show this behavior when the computer is in "sleep" mode.

☞ Jiggle your vision at a pulsating power LED and observe how the length of the dash relative to the length of the dark gap changes over time. Note that the *brightness* of the dashes doesn't change – the LED itself is not getting brighter or dimmer.

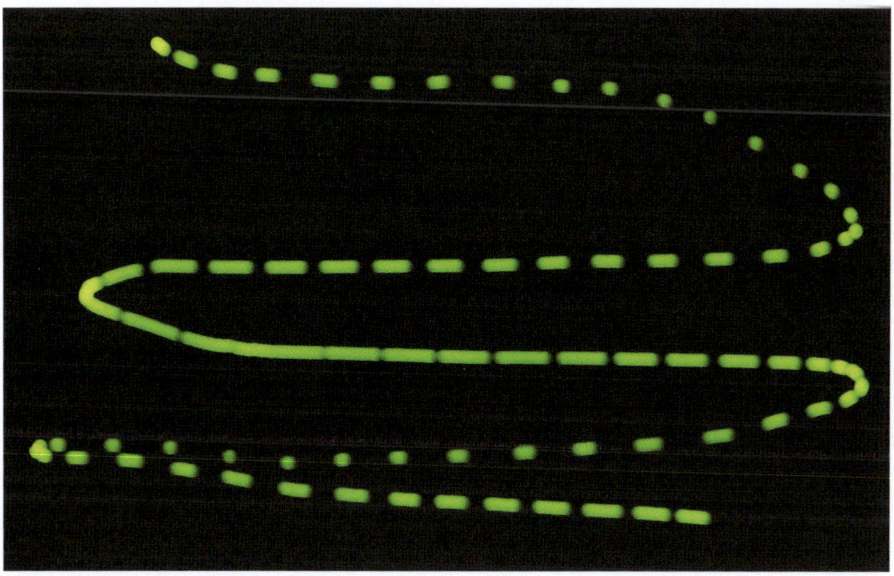

Figure 177: A trace of the power indicator LED on an electronic device (a Midiman / M-Audio MIDISPORT 4x4 MIDI interface). This LED appears to change brightness cyclically every two seconds or so. This exposure was made over 2.5 seconds, with a zigzag motion applied to the camera during the shot. As you follow the trace, you can see the ratio of light dashes to dark gaps is changing all the time. Small green dots with long gaps produce the dimmest illumination; long green dashes with short gaps produce the brightest. It looks like the image is out of focus: in fact, the focus is fine, but the illumination at the edges of the LED is soft.

Indicator LEDs

Indicator LEDs on electronic equipment are used to show the state of switches or some other status of the device. For example, a digital piano might have an LED above the "Reverb" button to show you whether the reverb effect is on or off.

Review the previous section on power LEDs (which starts on page 90) so that you can recognize LEDs that are driven in different ways: DC, AC and strobed. You may find any of these kinds when vision-jiggling at a device with multiple indicator LEDs.

The most interesting thing about a device with many indicator LEDs is that they may all be strobed *at different times*. Find a pair of LEDs that are side-by-side, or even a whole row of LEDs, and jiggle along an axis that's perpendicular to the line that runs through the pair or along the row. (Refer to the section on "Direction of jiggle" on page 19 for more details on how to select your jiggle axis correctly.) Look to see if all the dashes in your retinal trace are alongside each other, in which case the LEDs are strobing at the same time, or if they are misaligned, which means that they are strobing at different times. Quite often, the latter will be the case. Why?

The answer is that there's often just one circuit that controls all the indicator LEDs, and it maintains a list of which ones are on and which are off. Then it goes around in a loop applying a short burst of current to LEDs that should be on, and skipping over those that should be off. To the human observer, it looks like all the LEDs that are in the "on" state are steadily illuminated; but in fact it's possible that only one LED is actually illuminated at any time!

> ☞ On some devices, there may even be more than one circuit driving the indicator LEDs. See if you can figure out which groups of LEDs are driven by each circuit. (This may be hard to do.)

Note that numeric displays are covered on page 79. Also, see page 45 to learn about indicator LEDs and video cameras.

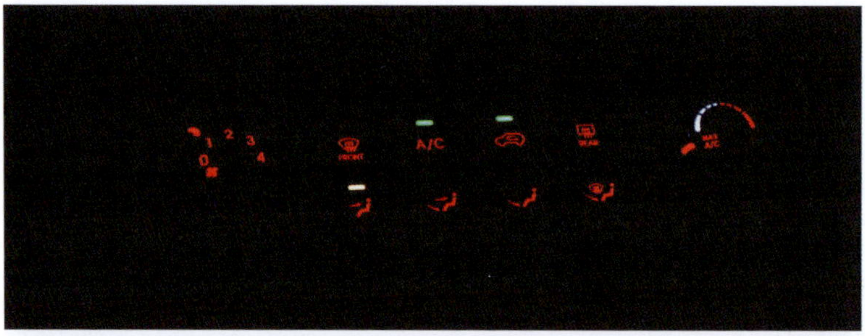

Figure 178: The climate control console of a 2011 Kia Sportage.

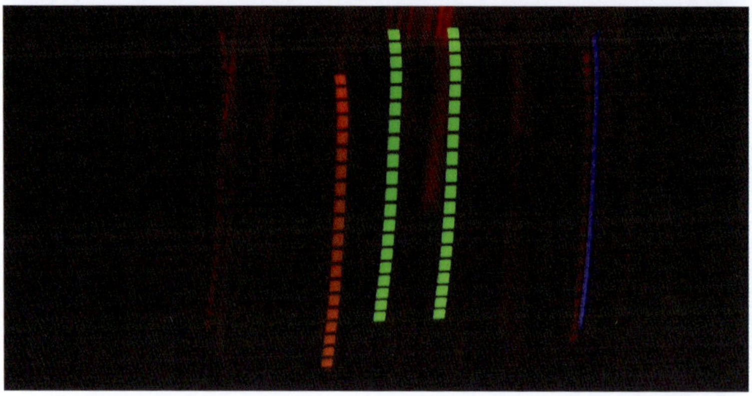

Figure 179: A photographic trace of the climate control console shown in the previous figure. It's clear that the indicator LEDs are flashing on and off rapidly. Note that the two green traces appear identical, which shows that they are flashing at the same time. This is not always true for other implementations. (1/10 sec)

Figure 180: An item of professional audio equipment in a rack. We're interested in the bottom row of red LEDs, which are all illuminated.

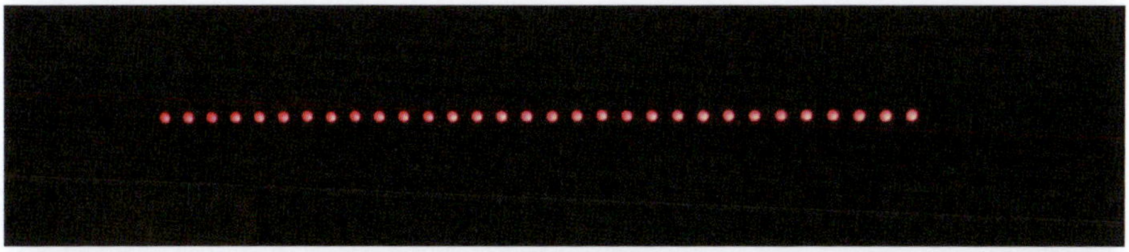

Figure 181: An under-exposed photograph of the same equipment, which makes it clear that the red LEDs are lit.

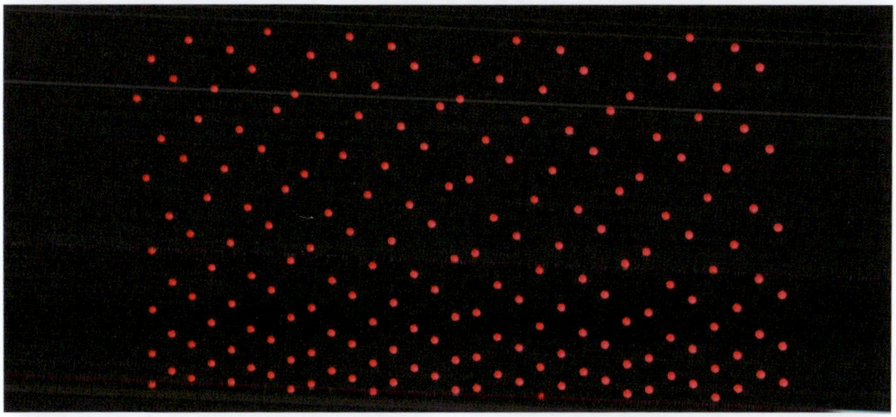

Figure 182: A vertically-swept trace of the LEDs. Each column of dots records the behavior of a single LED. Firstly, it's clear that none of the LEDs is DC-powered: they are flashed briefly with long dark periods between each flash. Secondly, the patterns of dots do not line up horizontally, so the flashes occur at different times for different LEDs. (1/10 sec)

Figure 183: *(left)* A digital audio mixing console.

Figure 184: *(right)* A photographic trace of the mixing console. All the LEDs are time-varying, including the two-digit red numeric display. (1/10 sec)

Digital projectors

Digital projectors are found more often in office environments than in the home; but some home theater installations use projectors. Projectors of this kind are usually based on LCD technology, though some use a digitally controllable reflection technology called DLP. They almost always use bright incandescent lamps for generating the light that's needed for projection, though some portable projectors are now using LEDs as a light source, though they are less bright than the ones with light bulbs. You may think that since the lamp is incandescent there would not be any interesting time-varying behavior to observe; but with of this kind of projector there are at least two interesting things you may see.

Many of these projectors operate by splitting the light into the three (additive) primary colors: red, green and blue. Each color of light passes through an LCD imager, which passes or blocks light according to the pixels on the computer's desktop. Some projectors achieve the splitting by passing three filters in front of the bulb: a red one, a green one and a blue one, one after the other, such as on a rotating "color wheel". When the red filter is in front of the bulb, the LCD imager shows the "red channel" of the computer's desktop. Then, when the green filter is in front of the bulb, the LCD imager shows the "green channel", and again for blue. I won't explain exactly what's going on here, or why it's done this way; it's outside the scope of this book.

Figure 185: A typical office projector.

During a meeting at which a projector is used, you may see transient colors appearing on the screen when you move your gaze from, say, the screen to a person. This is less clear if the screen is largely filled with white; but quite obvious for small light areas (such as text) on dark backgrounds.

For the observer of time-varying light sources, this type of projection is very interesting and rewarding. It's called "field-sequential" projection, and essentially the image projected on the screen is red, then green, then blue (the order may be different, depending on the projector). Usually, you don't see the sequential colors because they are happening so fast that the eye blends them together and sees the intended color for each pixel. But if you jiggle your eyeballs or use a mirror, then the sequential colors become very clear.

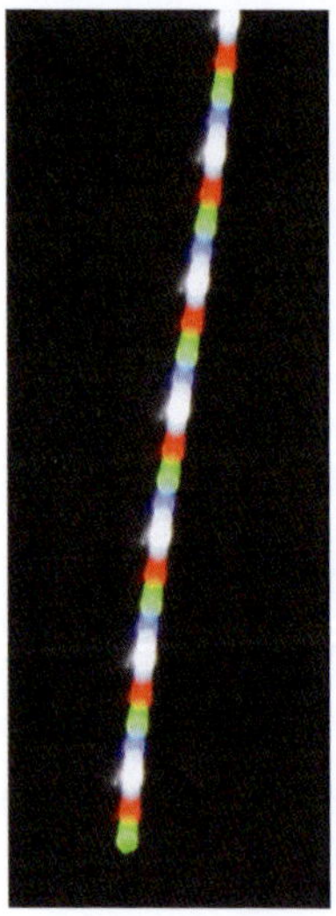

Figure 186: *(left)* This image is a trace captured by photographing a projector from the front. The round shape of the images in the trace is due to the round aperture of the projecting lens. The colored trace clearly shows a cycle of red - green - blue - white, which is typical for this type of projector. (1/10 sec)

Figure 187: *(right)* Projecting an image which consists of a white bar on the screen (see Figure 168) makes a good target for vision jiggling. This is the striking result, captured as a photographic trace. (1/10 sec)

Some field-sequential projectors also have a white period in the output, so it's common to see red, green, blue and white in the trace.

If possible, fill the screen with an image consisting of a white dot on a black background. The dot should be a couple of inches across when projected on the screen. Jiggle at it, and see if you can see a trail of strongly colored red, green and blue dots, or a continuous trace with a regular pattern of red, green and blue sections. If so, this is a sequential field projector. A black image with a white bar also works well and was used for Figure 187.

If possible, an even better way to see this is to stand in front of the projector and look back towards the lens. However, it's best to stand to one side so that you are just outside of the projected rectangle; otherwise it's likely to be too bright.

With a little practice, it's easy to see the colored fields even if the screen is showing something like a spreadsheet, which fills the screen. The more contrast there is in the content on the screen, the easier it will be to see the colored fields.

Another way to separate the colors is to wave something white or pale-colored in front of the projector. This dramatic photographic trace shows a hand being waved:

Figure 188: An image of a hand being waved in front of a field-sequential DLP projector. The trace at top left clearly shows a blue - red - green - white pattern.

And now for the second thing to watch for. Sometimes you will see a strong trail of dashes of white light when you look at the back of a digital projector. I have already indicated in this book that incandescent lamps don't vary significantly over time (see page 69 for an explanation). So what's happening?

The answer is simple: almost all projectors have one or more cooling fans which help to remove the heat generated by the lamp. If this fan is at the back of the projector, it's often possible to be seeing the bulb *through* the fan from your seat in the audience. The fan blades are moving too fast to see; but since they are passing in front of the lamp, the light from the lamp is being interrupted, and this produces a very strong time-varying behavior.

If you find yourself behind a digital projector, and if you can see white light escaping from the back of it, jiggle at it to see if it's strongly time-varying, i.e. flashing very fast. If it is, you are looking at the lamp through the fan. You may have to change where you are standing in order to have the fan between you and the lamp.

Strictly speaking, this observation falls into the "time-varying reflections and interruptions" category. More examples of this are given in the section on page 147.

Figure 189: *(left)* The rear side of a projector. Some of the light from the bulb escapes through the rear vent. A fan draws air through the projector to carry heat away. The fan's blades interrupt the light.

Figure 190: *(right)* A trace of the projector shown in the previous figure. The patches of light are slanted due to the fact that it takes time for the blades to move past the vent. The narrow vertical dark bar corresponds to the dark vertical support in the vent. Note that in this case the light that escapes the projector shows a periodic pastel coloration. This may be due to reflections from the color wheel, but I haven't proved this. (1/10 sec)

Appliances

Many appliances have the same types of light sources that are described in the chapter on consumer electronics, so please refer to that chapter for details. However, in this section we discuss the humble neon indicator, as seen on electric heaters, power strips and ovens, to give a few examples.

Anything with a neon power indicator

Neon indicators, which are typically red or orange, are probably the easiest type of time-varying light source to observe, and they make a great subject for practicing your mirror-jiggling or eyeball-jiggling skills. As I shared in the introduction to this book, it was a neon indicator that got me started.

One advantage of a neon indicator is that it's often attached to a piece of equipment that is easy to move around. The most accessible example is a simple electrical power strip.

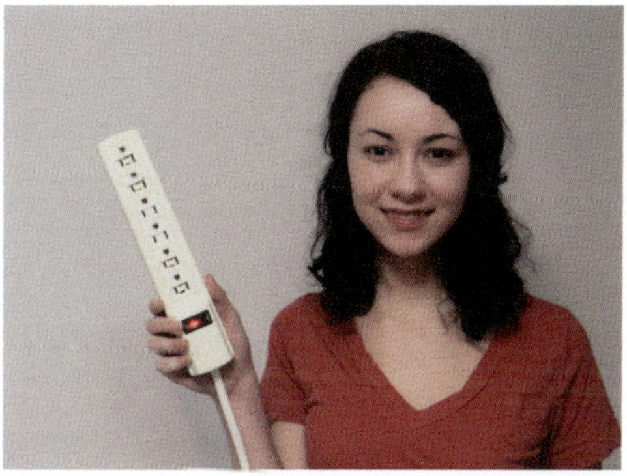

Figure 191: Many electrical power strips incorporate a neon indicator to show the presence of electrical power.

If you really need to be convinced of this whole retinal trace thing, go into a dark room and ask a friend to wave a plugged-in power strip around in front of you. You will definitely see traces.

Now, since most appliances don't jump around like this, it's up to you to move your eyes (or a mirror) to cause the moving of the image across your retina to generate the trail that allows you to unravel the light source's time-varying behavior. Refer to the early sections of this book to understand more.

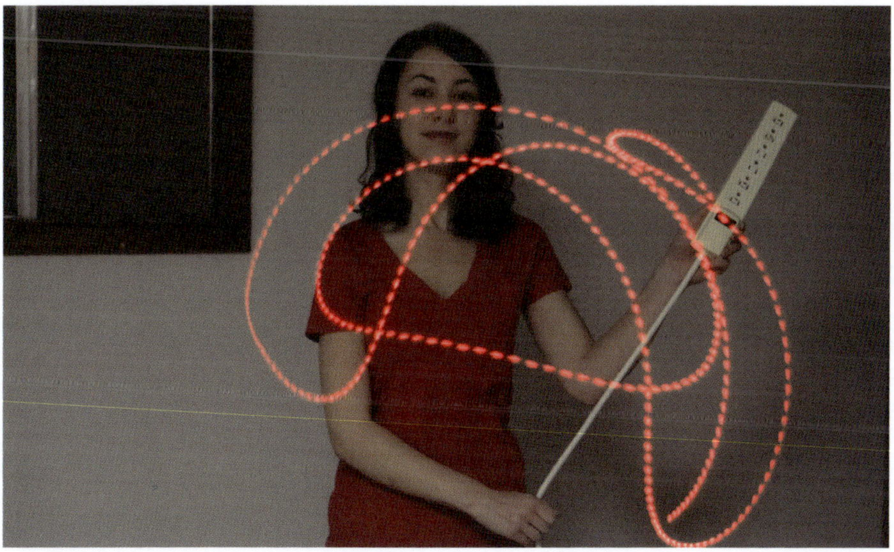

Figure 192: A photographic trace of a power strip being waved in front of the camera (also seen in Figure 9, where you can read details of how this photograph was taken).

Neon indicators have been used on electrical equipment since the 1920s, and are still probably the cheapest way to show that an appliance is switched on, especially if that appliance has no electronic circuits which could drive a light-emitting diode (LED). Most neon indicators are red or orange – after all, neon gas emits an orange-red color when it's stimulated by electrical discharge. There are other colors too, such as green; and while these are also typically called neon indicators, they have others gases in them.

Neon indicators flash because they are driven by the alternating current (AC) electrical supply in your home or office. Some indicators will light up on just one half of the AC cycle, giving a flash rate of 50Hz or 60Hz (depending on the frequency of the electrical supply in the country where you live). Others light up on both halves of the cycle, so the flash frequency will be twice as fast (100Hz or 120Hz). I expect that with practice, you'll be able to tell by vision jiggling whether they are 50/60Hz or 100/120Hz. (Imagine... a first date... "See that neon indicator on that refrigerator over there?" ...a suave look on your face... "It's illuminating on both halves of the AC cycle!" Your date is duly impressed.)

In a home or office setting, all outlets will be connected to a single phase of the electrical supply, so don't look for phase relationships as you would for street-lights (see page 51). In a factory, there may well be neon indicators on different mains phases, which might be interesting to observe, though I would expect the location to be brightly lit which would make observation difficult.

While writing this book I observed a power distribution unit designed for protecting a computer and its peripherals from power surges. It's pictured here:

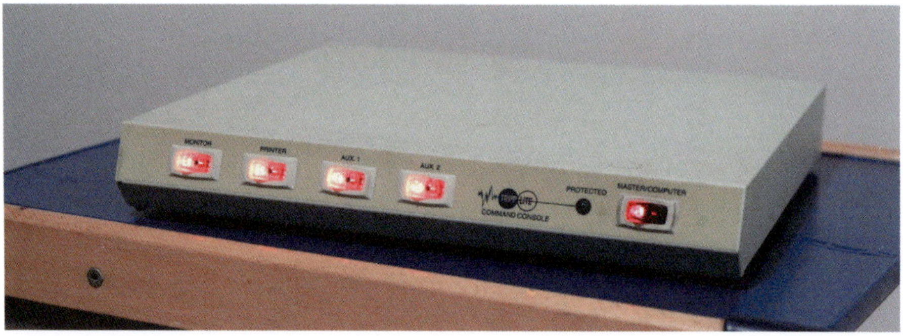

Figure 193: A power conditioner for computer equipment, with five neon indicator lamps.

It has a row of five neon indicators – one for the main switch, and one for each of the four circuits it controls. I jiggled my eyeballs briefly, expecting to see all the lights flashing in phase, and indeed I was right. However, I also noticed that the red dashes in the five traces in my retinal image were *not* the same length, but differed slightly from indicator to indicator:

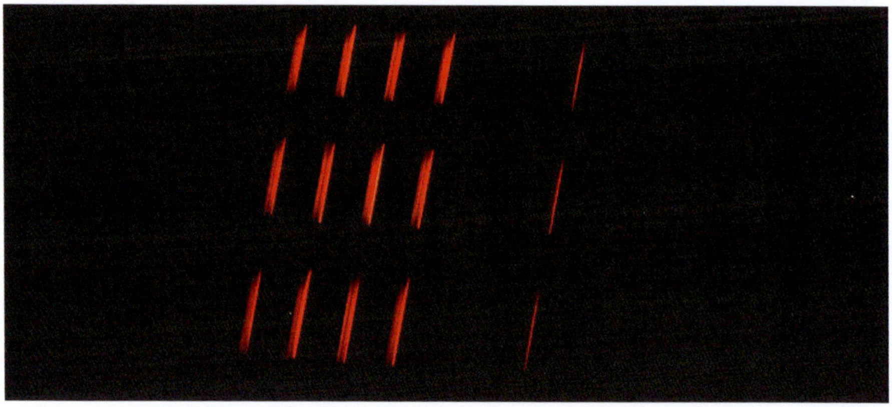

Figure 194: A photographic trace of the device pictured above. Note that the dashes are not all the same length, and that longer ones tend to start earlier and finish later than the shorter ones. (1/10 sec)

In fact, on closer inspection, the longer dashes were longer *at each end* than the shorter dashes. The conclusion: not all neon indicators light up at the same voltage; there's some variance from indicator to indicator. The following diagram illustrates this effect.

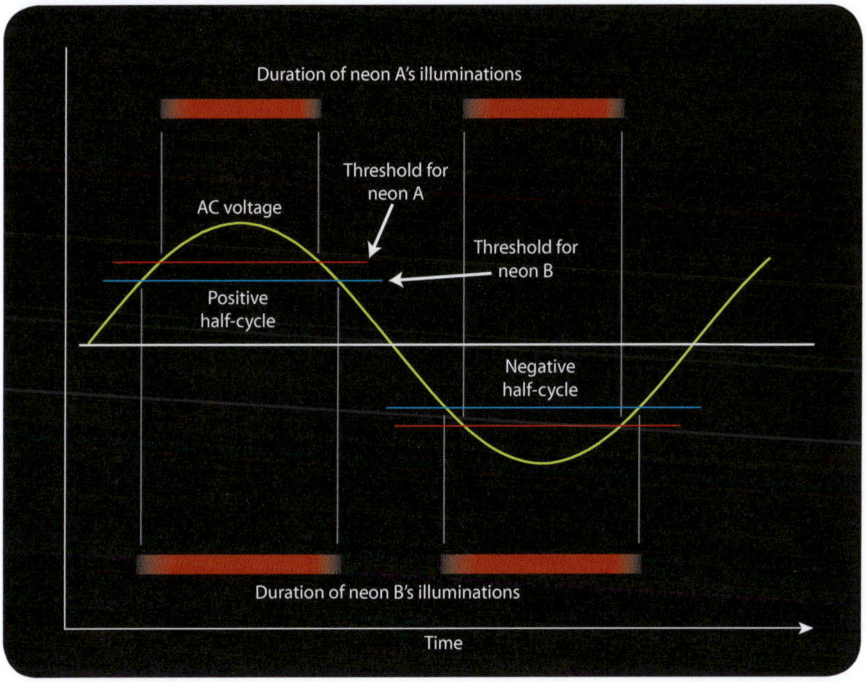

Figure 195: A diagram illustrating how different neon indicators may show different durations for their lit-up state. Imagine two neon indicators, neon A and neon B. Let's suppose that neon A has a higher voltage threshold than neon B, which means that the AC voltage must rise to a higher level (positive or negative) before neon A will light up, compared to the level needed for neon B to light up. The red lines represents neon A's "light up" threshold voltage; the blue lines are for neon B. Due to the sinusoidal shape of the voltage level over time, neon A will light up for a shorter duration than will neon B.

Next time you're in a dark room, at home or in a hotel, say, and there is a neon indicator in the room (on a fridge, wall-mounted air-conditioning unit or an electrical power strip), practice your eyeball-jiggling. Neon indicators are easy to observe and quite rewarding.

Signage

Illuminated signs are a feature of urban and highway scenery. Many signs, though not all, are time-varying. Some of them are among the most rewarding subjects for observation.

Neon signs

Neon tubes are probably the archetype of electric signage, especially in the United States. Introduced in the 1920s, they typify the brightest and brashest form of sign illumination, and are present everywhere there are stores or restaurants. For the vision jiggler, certain neon tubes are especially interesting as they show time-varying behaviors that you won't see in any other type of lighting.

Figure 196: *(left)* A neon sign.

Figure 197: *(right)* A photographic trace of the same sign. Two or three effects are evident, which are discussed in the text and ensuing figures. (1/10 sec)

True neon tubes are reddish orange: this is the color produced when electricity passes through ionized neon gas. Other tubes, which are also called "neon", are found in many different colors, including green, blue, purple, yellow and white. These are not usually filled with neon gas, but with argon or krypton, and in some cases with mercury vapor, which emits ultra-violet light.

Tubes with mercury are coated on the inside of the glass with fluorescent chemicals, which absorb the ultra-violet light and re-emit it at visible wavelengths, giving rise to the variety of colors seen in "neon" signage. Coated tubes like this operate in a similar way to fluorescent lighting; but due to the much higher frequency of the electrical power supply (at least in some cases), they are often not noticeably time-varying and are therefore not very interesting to the observer of time-varying light sources. Therefore, this section focuses on true (red) neon tubes, which are very interesting indeed.

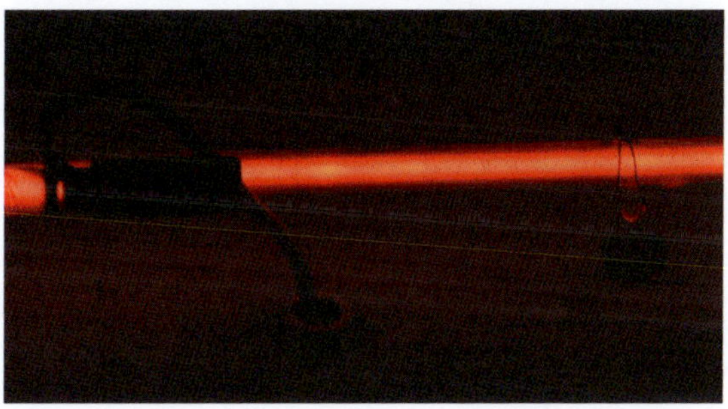

Figure 198: A neon tube that is not coated with a fluorescent material. Note how the brightest region is along the axis of the tube. This is by far the best type of neon tube to observe for time-varying behavior. Also note that the brightest line has "knots" or "clumps" of brighter gas. These variations give rise to amazing visual detail when observed with vision jiggling.

The first thing to consider is how to get the best experience when observing neon tubes. First, make sure it truly is a neon tube, and not a mercury/fluorescent one. If it's a bright reddish orange, then it's likely you're looking at neon. Another clue is if the light appears to be brightest in the middle of the tube, then it's the gas that's glowing, and not a fluorescent coating. Fluorescent coatings will appear equally bright across the whole diameter of the tube, since they're deposited onto the inside surface of the glass.

As with any time-varying light sources, the best results are seen when there are no other light sources present which could confuse what you see. Therefore, the very best type of neon sign to observe is a single, straight-line tube. These are often found as a decorative feature on a building.

Figure 199: A blue "neon" sign which shows essentially no brightness variation *across* the tubes. This shows some 120Hz variation to the jiggled eye, but the high-frequency behavior is impossible to see due to the persistence of brightness of the fluorescent material.

Figure 200: A restaurant with red neon lighting along the top edge of the building.

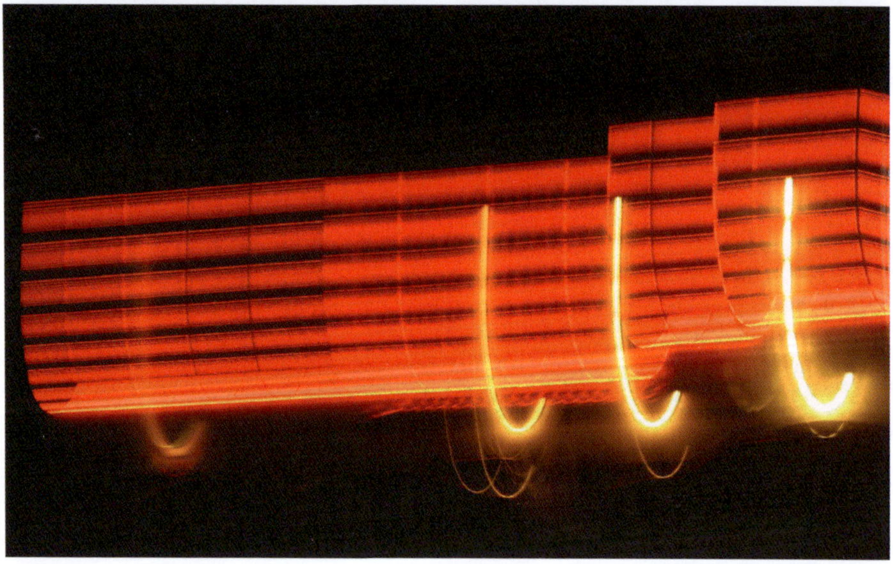

Figure 201: A photographic trace of part of the building. There is a wealth of information in this image! The rounded "swoop" at the bottom of the image is a consequence of the motion of the camera during the 1/10 second of the exposure. The large dark bands correspond to the 60Hz cycle of the AC electrical supply; there are actually 11 or 12 major red bands in this image, which means that this bright/dark variation happens 120 times per second (120Hz), which is twice the 60Hz mains frequency. This shows that the tubes are powered on both halves of the AC cycle. More detail is also evident: The bright bands are comprised of many tighter bands of red light; the left-most tube shows around fifteen such bands in each major band. This would indicate that the tube is flashing 1,800 times per second! Notice that the pattern and timing of the left-most tube is different from the other tubes: this indicates that the electrical circuit that drives it is separate from the others. However, the dark bars from the left-most and adjacent tubes line up: this indicates that they're on the same phase of the electrical supply. The right-hand tubes' dark bands don't line up; but this is due to the relative heights of the tubes, because of the shape of the building. It's clear on inspection that *all* the tubes are on the same phase of the electrical supply, if you compensate for the vertical displacement. Finally, there's an amazing effect which is not visible here. See the next figure for greater detail.

Remembering what we read on page 19 ("Direction of jiggle"), you will need to jiggle along an axis that's perpendicular to the neon tube. Therefore, if the tube is horizontal, you should jiggle up and down; if it's vertical, you will need to jiggle horizontally. This is because you're aiming to "paint" a band on your retina using the neon tube as a broad brush. If you jiggle *along* the axis of the tube, the tube itself will obliterate the image left on your retina, and you won't see anything interesting.

Figure 202: A photographic trace taken from a position much closer to the building. The detail is much clearer, and it's possible to see a "rope-like" or clumped effect on the images of the tubes. This is due to individual knots of light travelling along the tubes. In fact you can see that the "rope" is shifted horizontally between images of the tube (which merge together so it's not possible to see individual images of the tube), giving a "diagonal" look to the fine variations in brightness. Since images in the vertical stack are captured at consecutive times, this confirms that the flashes are travelling along the tube. The upper half of each major band shows the diagonal effect clearly. Notice how the *direction* of the diagonal effect alternates between the bands; one band may be "up and to the left"; the bands above and below it are "up and to the right". This supports the conclusion that both half-cycles of the AC supply (the positive half and the negative half") are used to power the tubes: the knots of light travel alternately one way then the other along the tubes. (1/10 sec)

Now observe the neon tube, looking for time-varying effects. There should be at least three effects going on:

- a slow variation in brightness, arising from the 50Hz or 60Hz AC electrical supply;
- a much faster variation in brightness, appearing as sharply-defined lines;
- and a pervasive observation of small, irregular, diagonal lines of brightness.

Let me explain these effects:

A neon tube typically pulses on and off as the voltage of the electrical mains supply rises and falls during the AC cycle. In the USA, this is sixty cycles per second (60Hz); in Europe it's 50Hz. Many neon signs are driven through a transformer which changes the 110V (or 220V) electricity to 3000V or more, in order to achieve electrical discharges in the neon gas. Transformers operate at the same frequency and with the same kind of profile (sinusoidal, in case you were wondering) as the input electricity. This results in a periodic, gradual brightening and darkening of the neon sign. And since it's usual for both halves of the AC cycle to be used for powering the tube, the actual rate of these brightenings is 120 per second (see the caption for Figure 201).

The appearance of clearly-defined lines of light comes about because when the voltage is just starting to rise during the AC cycle, the tube emits light in discrete flashes. This is because the voltage is not yet high enough to power the essentially continuous emission of light that comes soon after this phase. The discrete flashes are separated in time and therefore appear as lines (properly: images of the tube) with dark space around them. Figure 202 shows one such line very clearly; note too that just above that line is a series of closer-spaced lines which soon merge into continuous illumination. This shows that as the voltage of the AC supply continues to rise, the tube can flash more frequently since there's more energy available. Before long, the flashes merge into a continuous event which fades only when the voltage falls from its positive or negative peak.

While that's quite interesting, the *real* prize is the "diagonal" activity that's going on (see Figure 202). It's clear that there's detail in the red areas, with a random pattern that has a predominant diagonal character to it. What's happening?

The answer is that knots or pulses of brightness are traveling along the tube. Many of these pulses are present at any instant, like wagons in a train. Since these pulses are traveling *along* the tube and your eyes are sweeping past the tube in a *perpendicular* direction, what you will see is a *diagonal* effect. This is because your retina is moving, say, *down*, while the sparks are moving *across*. The combination of these two movements results in a diagonal appearance.

When vision-jiggling at a neon tube, this rippled effect is very easy to see. It may be harder to discern that there is a diagonal bias to the ripples, and very difficult to observe that the diagonal bias alternates one way then the other. As the photograph shows, there is some irregularity, so the patterns aren't always very clear. Also, the contrast between the light and dark regions in the tube may be quite low, resulting in the effect being subtle. But persevere, and you'll see it.

> One interesting effect that you might see is that even when two tubes are in line with each other, their dark bands may not line up. This is due to the tubes being connected to different phases of the electrical supply. Figure 203 and Figure 204 illustrate this effect.

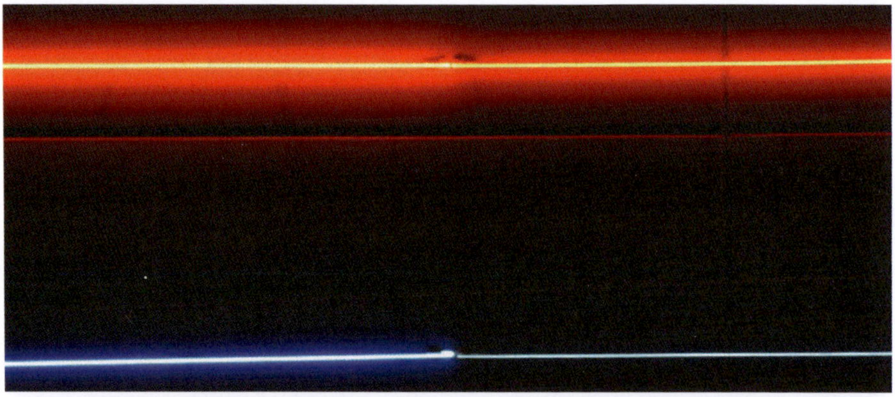

Figure 203: A photograph of two red neon tubes and two blue discharge tubes on the side of the American Furniture Warehouse building in Fort Collins, CO.

Figure 204: A photographic trace of the same tubes. Notice that even though the red tubes are physically in one continuous line, their patterns in the trace are displaced. This shows that the tubes are powered from different phases of the electrical supply. In fact, if you measure the displacement, it will always be exactly one third of the distance between the dark bands. This is due to the fact that there are *three* electrical phases; two are visible here. You might find a building where all three phases are present. For these tubes, the diagonal effect is very prominent; and, like in Figure 202, the direction alternates with each half-cycle of the AC power. (1/20 sec)

Finally, it's worth pointing out that neon signs and neon architectural decor make great subjects for time-varying art photography, as Figure 205 shows. Page 38 discusses the artistic possibilities of time-varying light source photography.

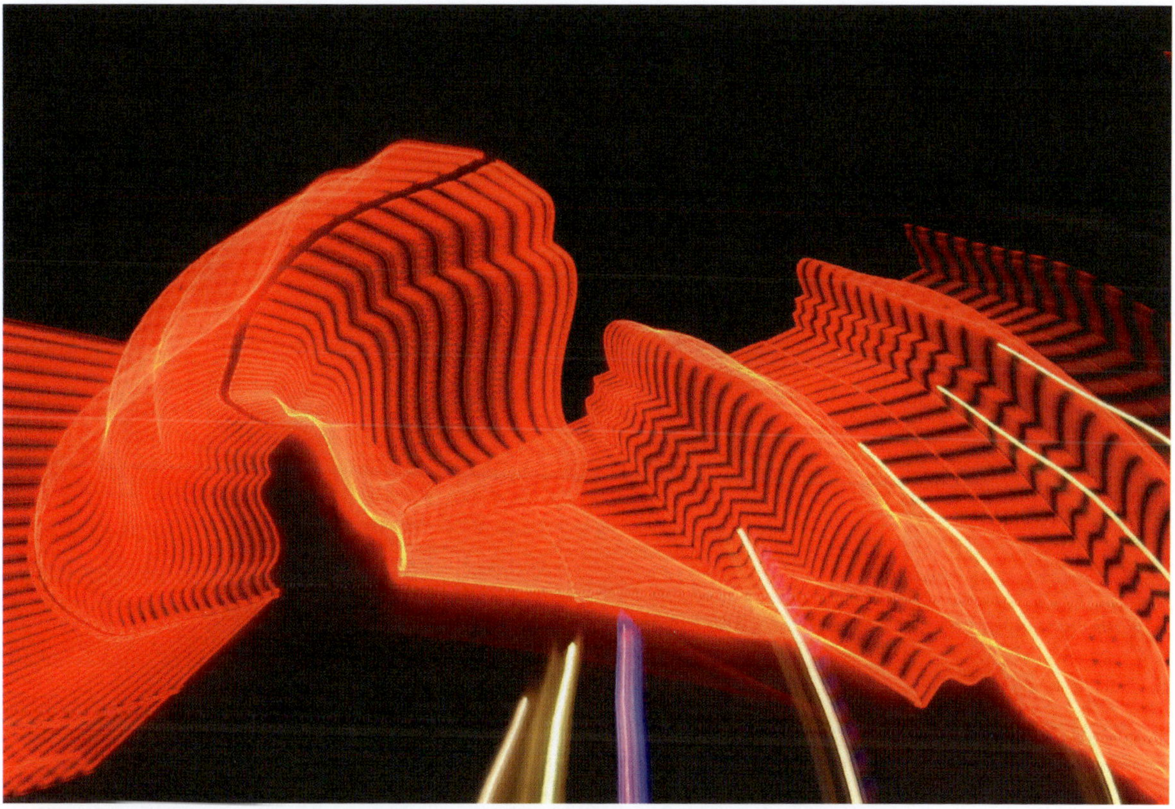

Figure 205: *(above)* Neon lighting, such as on this restaurant's roof, is a great source for photographic art. To achieve this shot, I rotated the camera and zoomed the lens during the exposure. (0.3 sec)

Figure 206: Here's the "reference photo" so that you can see the restaurant, and specifically the shape of the neon lighting. This photograph was not taken from the same viewpoint as the previous figure. Note that the neon tubes appear pink due to over-exposure, which was done to capture the scene. See page 33 for a discussion.

LED matrix displays

Matrix displays are very common, and are often seen at stores, gas stations, car and RV dealerships, and so on. They are also used on roads to warn of road work, traffic or weather conditions. There are almost as many types as there are individual displays, and they offer a great opportunity for the observer of time-varying light sources. By "matrix", I mean a rectangular two-dimensional array of lights which is controlled electronically to create text, symbols or images.

Most of these matrix signs use light emitting diodes (LEDs), though some older ones use incandescent light bulbs. Although many signs use a single color, such as red or orange, others can generate a range of colors by blending the light from different colored LEDs, most often red, green and blue. Some displays are not time-varying, with the "on" LEDs permanently on. These are typically very bright. Most displays, however, are time-varying, and are therefore interesting to the reader of this book.

Single-color LED matrix displays

Single color displays are the most common. It's usually easy to observe them at night, unless there are other lights nearby which might interfere with what you're seeing. Try masking the view (see page 20).

Figure 207: A sign outside a gas station. The green display is what we're interested in. Fortuitously, the red display that's just above the green display wasn't working (except for the dot), which makes our observation easier.

There are several primary types of time-varying behavior that you're likely to see:

- Steady-state, or DC powered. These signs are not flashing; all the segments or dots in the display are on all the time. These signs are often very bright.

Figure 208: A photographic trace of the green display (and the red dot) in the previous photograph. Note that the green trace is solid: there is no banding in it. This shows that the sign is DC-powered, or "always on". (1/10 sec)

- All the LEDs across the whole display strobe (flash) in unison. This would typically serve the purpose of controlling the brightness through modulation, much like LED tail lights on cars (see page 59).
- The LEDs flash in sequence, row by row as the controlling electronic circuit lights up all the LEDs in each row and then the next row. This not only controls the brightness, but also saves on electronic component cost by only needing to light up the "on" LEDs in one row at a time. The eye sees the display as if all "on" LEDs are on all the time; but the vision-jiggler would be able to see the true pattern of illumination. When observing this type of display, and assuming that the rows are illuminated in top to bottom order, you should be able to see that an upward sweep will stretch the letters vertically, and a downward sweep will squash them vertically or – if done fast enough – even invert the letters. (This is similar to the vertical single-sweep observations for a CRT television; see page 81). Also of interest is jiggling side-to-side: the text will appear to tilt.
- The LEDs flash in sequence, column by column. This is similar to the preceding case, and is simply a different implementation of the electronics. Vertical jiggling will be very interesting, as you will see the whole row of text skewed upwards or downwards.
- There may be other time-based patterns in which the LEDs on a given display are illuminated (e.g. Figure 218). Observe and try to figure out what's going on!

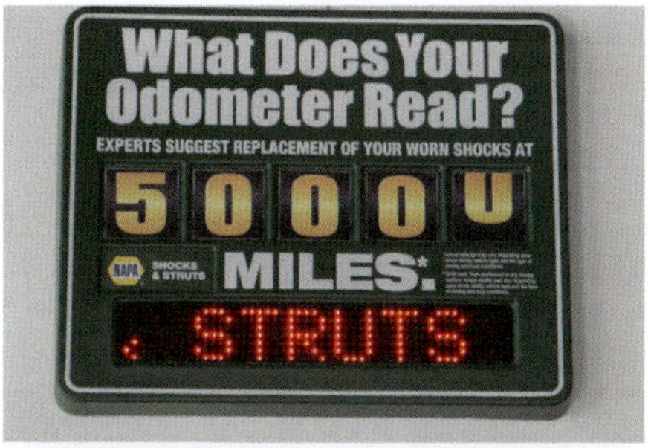

Figure 209: *(left)* A typical LED matrix display.

Figure 210: *(right)* Because of the time-varying nature of the matrix, it's easy to capture the display when not all the LEDs are illuminated during the time the shutter was open. This is a side-effect of a time-varying light source on photography; see page 35. (1/200 sec)

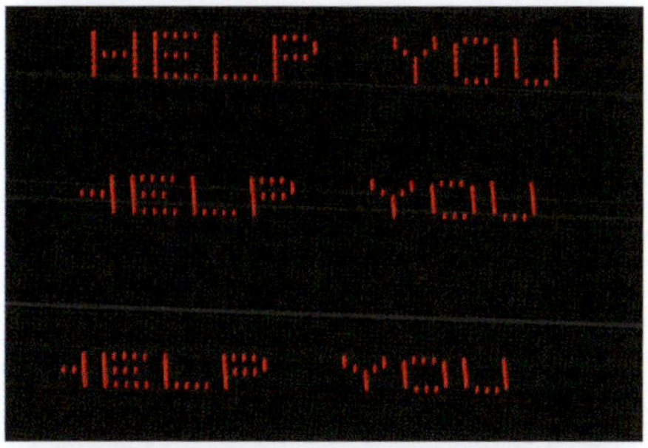

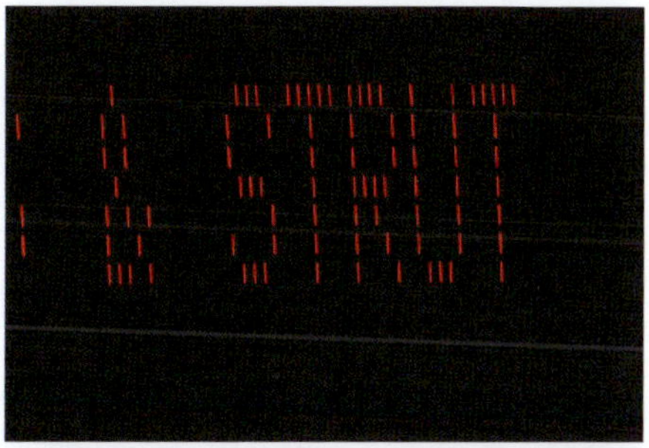

Figure 211: *(left)* A downward sweep compresses the text, which confirms that this display is illuminated row-by-row, top row first. (1/10 sec)

Figure 212: *(right)* Conversely, an upward sweep stretches the text vertically. (1/10 sec)

Figure 213: *(left)* Just as for a CRT TV, sweeping the camera to the left causes the text to tilt to the left. (1/10 sec)

Figure 214: *(right)* Sweeping to the right causes the text to tilt to the right. (1/10 sec)

Sometimes you will see a display with, say, red LEDs, on which a single line of text is scrolling across smoothly from right to left. This type of display is most often seen in store windows. Our local DMV office used to have one that encouraged you to be an organ donor; you know the kind I mean.

Figure 215: A temperature display on the wall of a bank building. The orange color of the LEDs has been washed out to yellow by the camera's sensor (see page 41).

Some of these displays show a very interesting phenomenon: while the text is moving, it appears to be gently slanting (almost as if it were italic text); but when the text stops moving it's suddenly standing upright. On closer inspection, you find that the LEDs are arranged in a square grid pattern, and that it would in fact be impossible for the letters to slant at the gentle angle that you saw when they were moving: the LEDs are simply not arranged that way. What's going on? (I'm not giving the answer here; but here are some clues: vision-jiggling will tell you in what sequence the LEDs are illuminated; and remember that when you're watching the smoothly-scrolling text, your eyes are following it smoothly...)

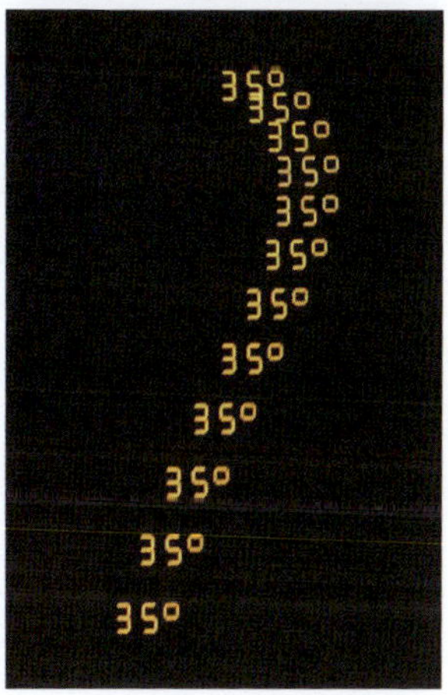

Figure 216: A photographic trace of the same display. It's very clear that all the LEDs in the entire display flash on and off simultaneously. The clarity of each image shows that the "on" time is *very* short; there's no smearing of the images as the camera was swept. In fact, this display is in reality dark almost all of the time! Twelve images in this 1/10 second exposure show that the flash rate is 120 Hz, which is very common in the United States; it suggests that the electronic circuit derives its timing from the 60 Hz AC electrical supply.

Once you have figured *that* out, what's *really* interesting is what you might see when you *don't* follow the moving text with your eyes but continue to watch it attentively: the text may *still* look like it's slanted! I would love to hear an explanation of the effect! (I believe it's a result of retinal motion processing.)

Figure 217: *(left)* An LED matrix sign.

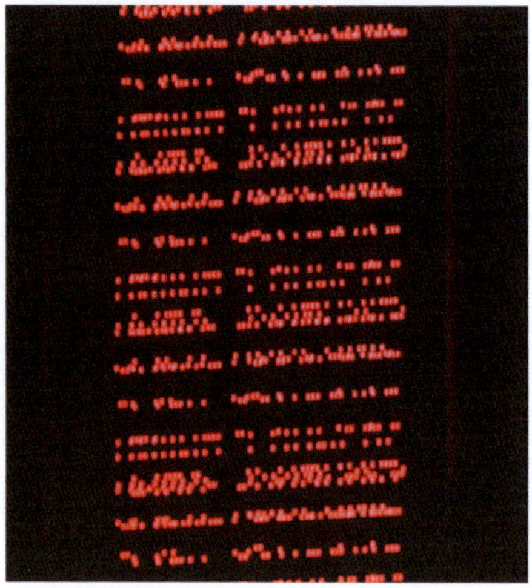

Figure 218: *(right)* A truly bizarre photographic trace obtained from the sign in the previous figure. I expect that by carefully inspecting this trace and the reference photograph it would be possible to determine the exact sequence in which the rows of pixels are illuminated, for clearly it's some row-by-row scheme. However, when observing it with eyeball- or mirror-jiggling, it's not going to be possible to do a detailed analysis. It is, however, every bit as amazing to observe as this trace would suggest. (1/10 sec)

Colored LED matrix displays

As with TVs and computer monitors, it's possible to present a wide range of colors to the eye by mixing three primary colors: red, green and blue. There are LED matrix displays which operate in this way; and there are some which have only two LED colors and deliver a narrower range of colors. In either case, it's pretty much a given that a multi-color LED matrix display will show interesting time-varying behavior.

The sequences I listed above, in the single-color display section, may also be observable with multi-color displays; but what's more interesting is that the different colored LEDs may light up at separate times in sequence. So, for example, all the red LEDs might flash, then the green, then the blue. By jiggling your vision, you will be able to separate out, at least partially, the different colors in the display. This is similar to the "field sequential" technique used in digital projectors (see page 93).

Such signs are often very bright, so it's helpful to observe them from a distance. This will also help with separating out the images in your retinal trace, as the size of the display itself can lead to it obscuring its own trace; observing it from further away means that it will appear smaller (see page 21).

Figure 219: *(left)* A color LED matrix sign.

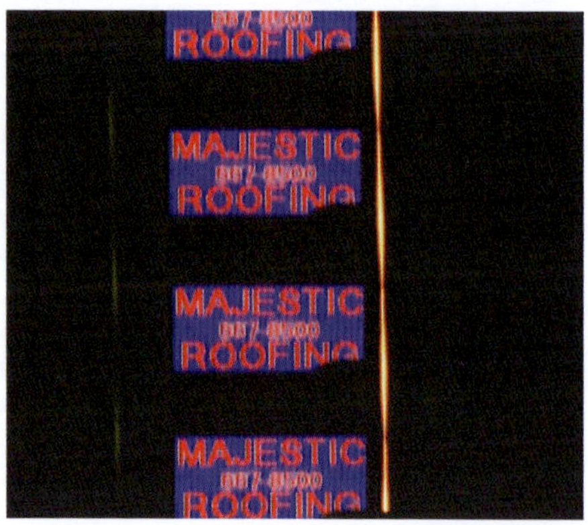

Figure 220: *(right)* In this case, *all* LEDs in the sign flash at the same instant. There are no separate red, green and blue phases. The images of the sign are not vertically stretched or smeared at all, which indicates that the LEDs are illuminated very briefly. This sign is dark much more than it's illuminated. Note the pressurized sodium light to the right. (1/10 sec)

Figure 221: *(left)* This photograph and the following five photographs were captured from a color LED sign outside an RV dealership. The image on the sign changed every few seconds through a cycle of three messages. Here is the first one.

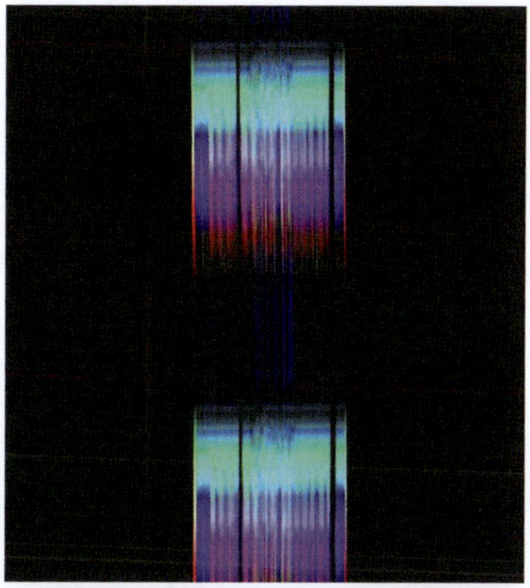

Figure 222: *(right)* A photographic trace of the sign when showing the "Visit KCRV.com" message. Although it's possible to see broad green, blue and red periods, there are complexities in the time-varying behavior which defy easy description. (1/10 sec)

Figure 223: *(left)* The second message on the sign.

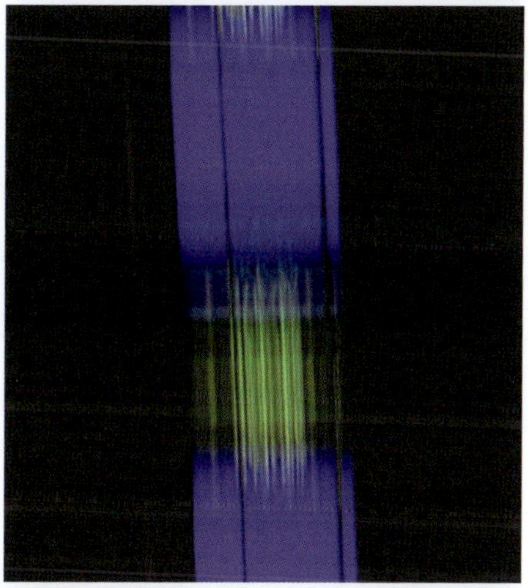

Figure 224: *(right)* A trace of the "Show prices here" message. Clearly, blue and green occur in alternation, but again there is deeper complexity in the timing. (1/10 sec)

Figure 225: *(left)* The third message.

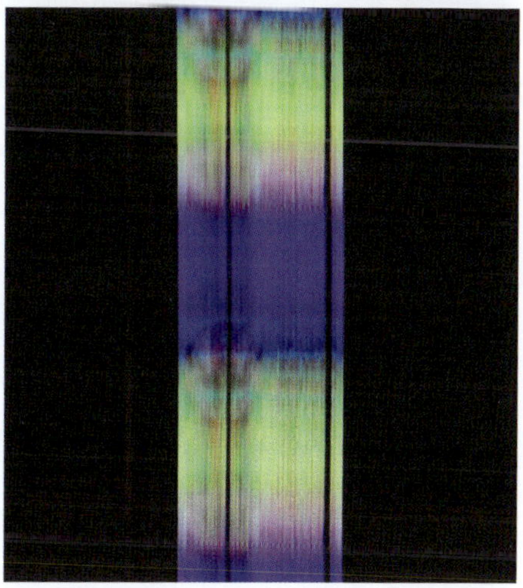

Figure 226: *(right)* A trace of the message in the preceding figure. It's amazing to see how the sign's time-varying behavior is so complex, with strong blue and yellow periods. Note that the yellow period indicates that both green and red LEDs are illuminated together. (1/10 sec)

Fluorescent signs

Many signs that you see outside stores are illuminated with fluorescent tube lighting, which is covered in detail on page 70. The sign is essentially a box with fluorescent tubes inside; the light shines through the faces of the box, on which the words and images are applied.

If you see such a sign, you will be able to tell if low-frequency fluorescent lighting is used by seeing if there are dull yellow bands in the retinal trace. (High frequency fluorescent lighting will show no time-varying behavior - see page 74.)

Figure 227: A commercial sign lit with fluorescent tubes.

Figure 228: A typical fluorescent-lit commercial sign with the front screen removed.

Figure 229: *(left)* A fluorescent-lit sign on a building.

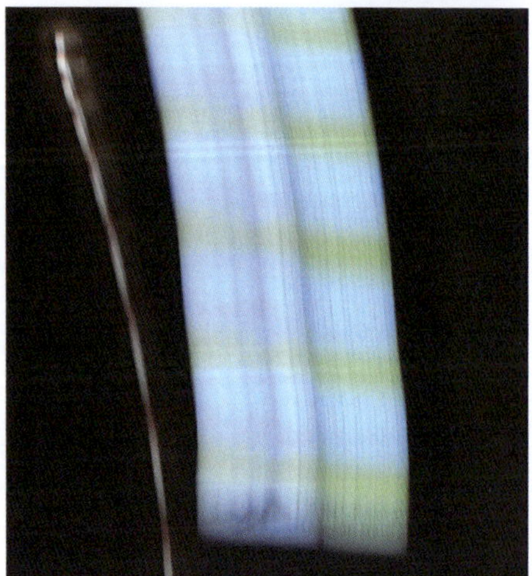

Figure 230: *(right)* A photographic trace of the sign in the previous photograph. The yellow bands are clearly visible, confirming that this sign is lit with traditional fluorescent tubes. (1/13 sec)

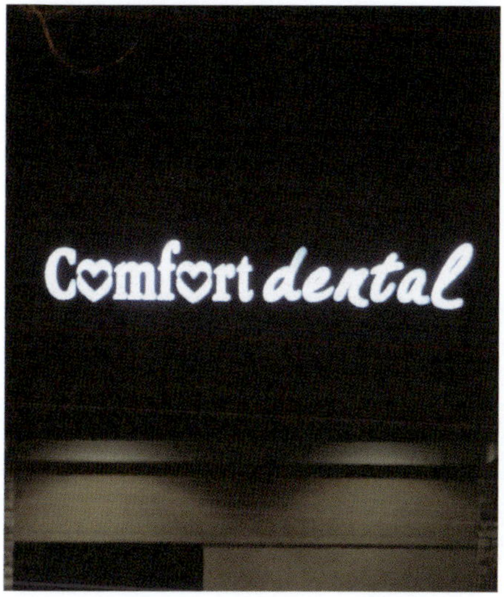

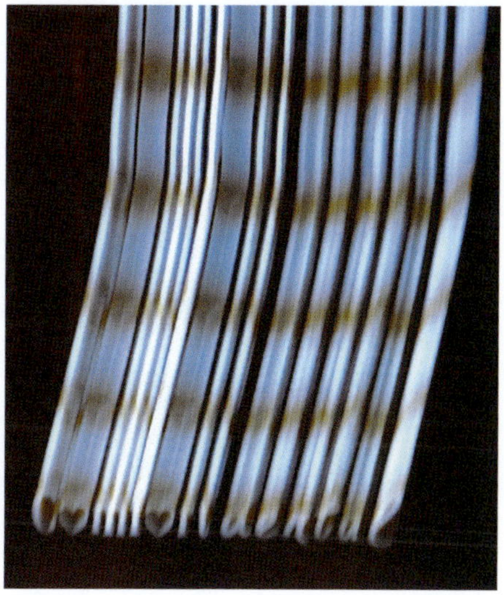

Figure 231: *(left)* A more complex sign...

Figure 232: *(right)* ...which is proven to be lit by fluorescent tubes. Again, the dark yellow (brown) bands are very apparent. (1/10 sec)

Incandescent signs (flashing)

We have seen many times in this book that incandescent lighting is not usually a time-varying light source, and so it's not very interesting for the vision-jiggler. Details are given on page 69.

Now, you might expect that light bulbs that are flashing rapidly, such as you might find at a carnival or in Las Vegas, would be interesting to observe; but sadly, this isn't usually the case. The flashing is simply too slow for the retinal trace to persist long enough to reveal the time-varying behavior. Due to the nature of incandescent lamps, it's impossible to flash them very fast, because they take a relatively long time to turn on and off.

You may have better luck with photography: I suggest a long exposure time, such as one second; and a small aperture to keep the image from being over-exposed.

Figure 233: Boring? (Photograph: BigThunderMtn)

Other time-varying light sources

Air traffic warning lights on radio masts and buildings

Lights on a transmission mast or other tall structure are a common sight; they are required by regulations to provide safety for low-flying air traffic which might otherwise fly into the structures. These lights are usually red, though white ones are also found. Some are incandescent and are not visually time-varying (see page 69). Some use high-intensity LEDs, and are usually DC-powered, which means that these also are not time-varying.

A third type is a xenon discharge light similar to what you would find in a photographic flash gun. These operate by emitting *very* short flashes of light separated by a relatively long period of darkness. Both red and white lights can be of this type; xenon emits a white flash, so the red lights have red glass filters over them.

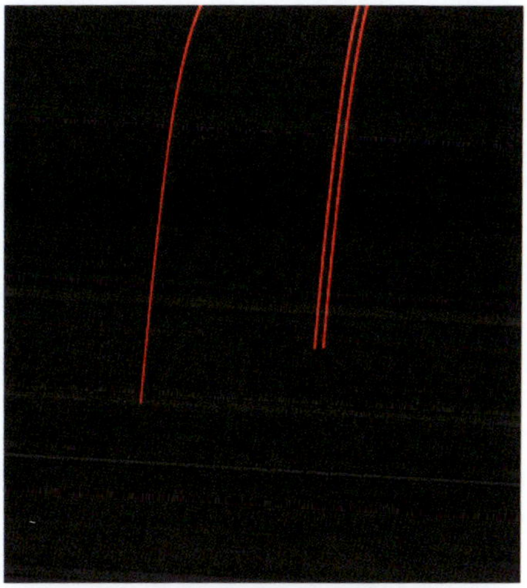

Figure 234: *(left)* Red aircraft warning lamps on a hotel building near an airport.

Figure 235: *(right)* The photographic trace of the hotel in the previous figure shows that the lamps are not time-varying. They are either incandescent (light bulbs) or LEDs driven by a steady DC current. (1/10 sec)

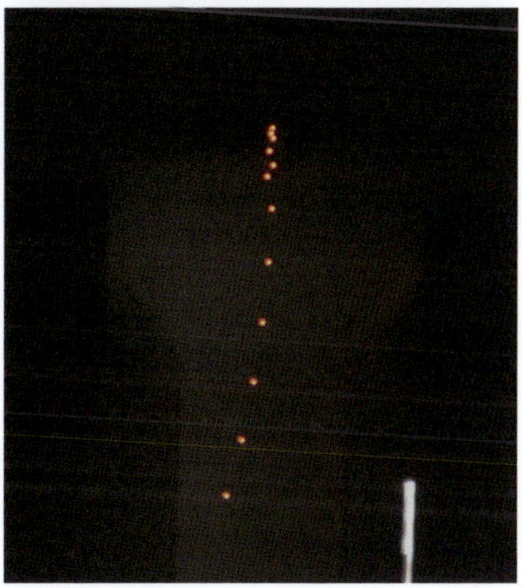

Figure 236: *(left)* A water tower with a red aircraft warning light which does not (apparently) flash.

Figure 237: *(right)* However, this photographic trace reveals that the aircraft warning light is flashing rapidly with very short flashes, which are typical of xenon discharge lamps. This is one of the easiest targets for the vision-jiggler, with the separate red flashes being very easy to see. This photograph was taken at a 1/10 of a second exposure time, and there are twelve flashes, which indicates that it's flashing at 120Hz. This is expected: it's a multiple of the 60Hz mains frequency, and shows that the circuit generates a flash for each half-cycle of the AC electrical supply.

Find a tall transmission mast at night and jiggle at it from a distance of at least a quarter of a mile. Since many masts have multiple lights, each at a different height up the mast, you may find that jiggling horizontally works better than jiggling vertically (see the section on "Direction of jiggle" on page 19). First, determine if the lights are steady-state or time-varying (note: there may be both types on a single mast).

Figure 238: *(left)* Two transmission masts against a sunset. Each mast has a single red beacon at the top and a group of four red beacons about half-way down.

Figure 239: (right) A photograph of the beacons on the two masts, zoomed in relative to the photo on the left. I have cheated a little: I enlarged each light in Photoshop to ensure that the lights will appear clearly in the printed form of this book.

Figure 240: A horizontally-swept trace of the beacons on the two masts. The top beacon on the left mast is clearly a xenon discharge lamp with very short flashes which are easily observed by vision jiggling. All other beacons are constantly on. The two top beacons flash on and off every couple of seconds, so I had to time this photograph carefully to include both of them. (0.4 sec)

Figure 241: A closer look at the xenon discharge beacon. Although the camera was swept rapidly, the individual images of the lamp do not appear horizontally stretched, supporting the observation that the flashes are extremely short-lived. There are 12 images in this 1/10 second exposure, which indicates that the beacon flashes 120 times per second when "on". This rate matches the electrical mains frequency (60Hz), with the lamp discharging on both half-cycles.

For discharge lights, note how spaced apart the flashes are. In fact, no matter how fast you jiggle, you will probably never see the flashes as dashes: this is because the flash is so short – less than millisecond – that it's not illuminated for long enough to paint a trace on your retina. If you think about it, this means that the lights are actually *dark* almost all the time! However, without vision-jiggling, you perceive them as being *on* all the time. Such is the effect of your persistence of vision (see page 11).

Finally, it's worth describing the most interesting aircraft warning lights I have ever seen, which were in Sweden. Near Gothenburg (Göteborg) is a transmission mast (the Brudaremossen mast) with three or four aircraft warning lights on it. When I saw it in the late '90s, each light consisted of a cluster of bright white lights with an armature rotating around it. This armature carried prisms or lenses, with the effect that each cluster was like a crazy lighthouse with maybe a dozen beams coming from it. The armature rotated quickly, so that the beams swept across your eyes very fast. It was certainly very eye-catching, which is presumably the aim of this design. Sadly, I did not get the opportunity to eyeball-jiggle at it; and for all I know, the lights may have been replaced with more modern technology. If it's still there, I would welcome a report from any Swedish readers.

Aircraft landing lights

All commercial aircraft have landing lights, which are illuminated when approaching the destination airport at night. In almost all cases, these lights are incandescent, emitting a steady beam with no time-varying behavior. However, it's common to see irregular variations in the brightness of the light, as seen from the ground. These variations are caused by irregularities in the density of the air that's between the aircraft and you. It's essentially the same effect as the atmospheric shimmering you can see on hot days; only in this case the only visible result is the variation in the landing light's brightness; and while this can be observed directly, it becomes very apparent when vision-jiggling.

Figure 242: A photograph of an airplane coming in for landing at Las Vegas airport. This aircraft has four landing lights.

Figure 243: A trace of the aircraft shown in the previous figure. There is no evidence of regular time-varying behavior, which is expected because these lights are incandescent. However, there are clear variations in brightness along the traces which result from density variations in the air between the observer and the aircraft (about seven miles in this case). Notice the short-duration flash of the wing-tip beacons near the bottom of the trace. These are xenon discharge lamps. (1/10 sec)

- Find a location a few miles from an airport at night and watch the airplanes as they approach. You may need to determine which direction they are coming from (they will be flying *into* the wind) so that you can see their landing lights.

- The British band SKY (a rock band comprised of classical musicians) recorded a track called *Watching the Aeroplanes*. You might want to put it on your MP3 player for this occasion.

Airport marker lights

Whenever I am on an airplane that takes off or lands at night, I enjoy looking at the large number of colored lights on the ground that mark features and zones on the runways and taxiways. Several colors are represented, notably blue, green, orange and red.

Older lights are incandescent, and show no time-varying behavior. Their colors are also a little washed out.

Newer lights are implemented with LEDs, and these are, in my experience, always flashing rapidly, and therefore a great subject for the vision-jiggling observer. In most cases, the flashing is a simple 50% on / 50% off cycle. However, other patterns do exist, though they may be hard to observe without taking photographs.

Thankfully, now that airline carriers are relaxing the rules regarding the use of electronic devices when taking off or landing, taking photographs of these lights is no longer the surreptitious activity it used to be!

LED marker lights are often brighter and nearly always more saturated (stronger color) than incandescent lights. At many airports, such as Denver International Airport, it's possible to observe both kinds.

At some airports, the runway itself is interesting to observe. For example, the runways at Albuquerque's Sunport airport have a line of red and white lights along the middle. All of these lights are flashing rapidly, *at the same time*. For the vision jiggler, the result is that the whole line appears multiple times in the trace.

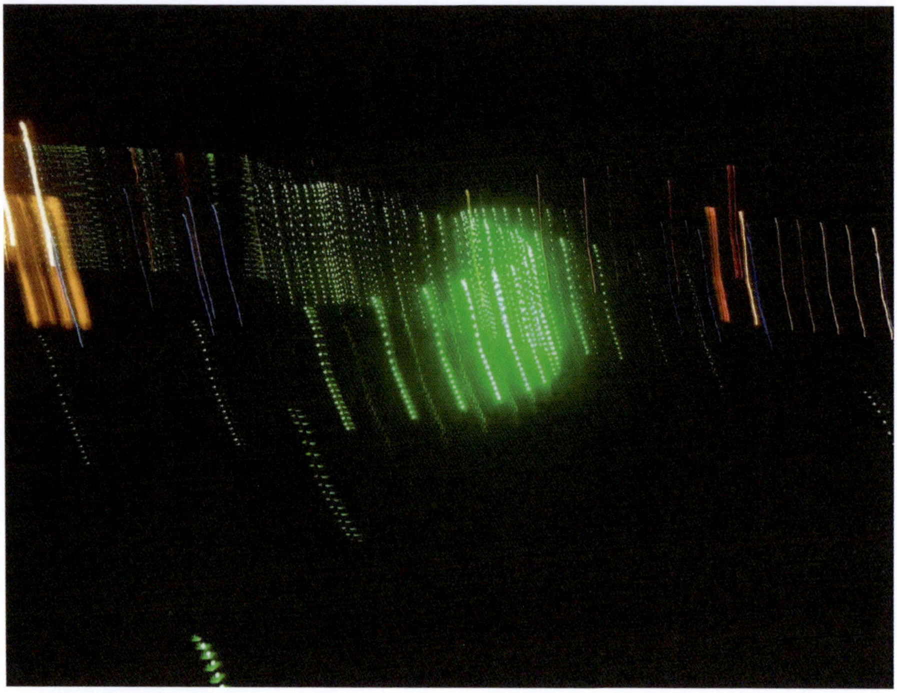

Figure 244: A remarkably festive photographic trace of green taxi-way marker lights at Denver International Airport. It's very clear that these lights are flashing rapidly. It is very easy to observe this with the naked eye. This photograph was captured with an LG G2 cellphone, and the shutter speed was not reported. However, since each trace has 20 dashes, I would judge the exposure time to be 1/3 second (assuming 60Hz flashing).

Christmas lights

Until recently, Christmas lights – while very pretty – were generally not very interesting to the vision-jiggler. This is because most lights were incandescent (little light bulbs), and we have seen many times in this book that incandescent lights are on the whole not time-varying (page 69), even when they are visibly flashing (page 111).

Figure 245: A house decorated with Christmas lights.

Figure 246: A trace of the house from the previous photo, which demonstrates that the lights are almost certainly incandescent, as there is no time-varying behavior (no flashing). The only trace that does show variations (hard to see) corresponds to the reflection of a street light in the rightmost of the group of three windows. (1/10 sec)

However, times have changed, and many Christmas lights are now comprised of light emitting diodes (LEDs), which are much more interesting.

LEDs used in Christmas lights are often driven directly from the AC electrical supply (through a resistor to limit the current), and as such they will behave in the same way as an AC power LED, which is discussed on page 90. In short, this means that the LEDs are on for 50% of the time (or a little less), and that the transitions from "off" to "on" and from "on" to "off" are gradual, not instantaneous. Some lights, however, are driven through a circuit which makes the on-to-off and off-to-on transitions more abrupt.

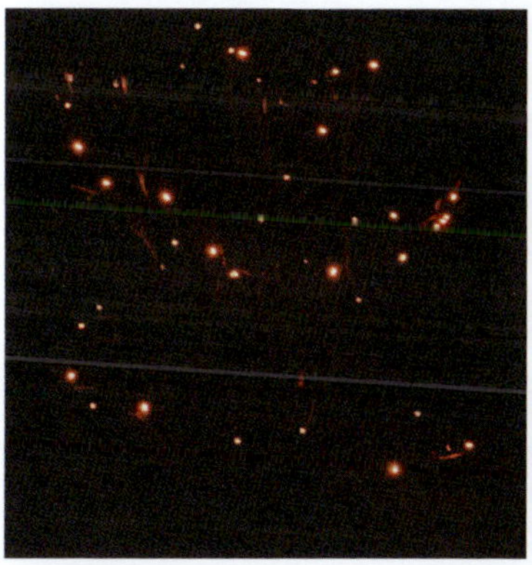

Figure 247: A small bush with a string of orange LED Christmas lights draped around it.

Figure 248: A photographic trace of the bush in the previous figure. In this case, I swept the camera sideways, not downwards. Note how every LED paints a clearly-marked series of dashes, with abrupt start and end points to each dash. Closer inspection shows that *every* trace begins with the same pattern of dashes and ends with the same pattern of dashes. This shows that (as one would expect) all the LEDs are flashing at the same times. (1/10 sec)

A characteristic of many Christmas lights is that all the LEDs in a string are strobing simultaneously, so it looks like the whole decorated tree or feature is flashing.

Figure 249: A row of LED-lit trees, taken from a distance. Note the lamps to the left of the left-most tree and the right of the right-most tree.

Figure 250: A photographic trace of the row of trees in the previous photograph. At this distance, the camera (or your eye) can sweep through a much larger angle than if you're close to the trees. The individual LEDs aren't visible; instead, the trees themselves prove to be flashing. Note how each "blob" of light is tree-shaped. By the way, we now see that the lamp on the left is incandescent (no time-varying behavior), but the one on the right is time-varying (possibly a discharge light). (1/10 sec)

It's common to find many LED-draped trees around shopping malls at Christmas time. Jiggle your vision and see how all the trees are flashing. Are you able to tell if they are all connected to the same phase of the electrical supply or not? (Note: don't forget to account for spatial alignment – see page 18 and page 50).

Other strings of lights show a different behavior: sections are lit up at different times. For example, the strings on the house pictured below are illuminated in alternating sections.

Figure 251: A house decorated with modern LED Christmas lights. My camera tends to record some white LEDs as blue.

Figure 252: A photographic trace of the house in the previous photograph. Look at the white lights which run along the top of the roof. You can see that there is an alternating pattern of light and dark sections, which shows that the string of LEDs is arranged in such a way that alternate sections are on and off at any time; and then a moment later the on sections switch off and the off sections switch on. Therefore, approximately half of the LEDs in this type of string are on at any time; the others are off. This is a very common type of decorative LED string, and can also be seen in multi-colored strings. (1/10 sec)

- Look at strings of Christmas lights along roofs and fences. Determine if they consist of incandescent bulbs or LEDs.

- If they are LEDs, do all the LEDs flash at the same time, or are they arranged in alternating sections?

Figure 253: A domestic front porch festooned with LED Christmas lights.

Figure 254: A photographic trace of the Christmas lights in the previous figure. The illuminated doorbell button is the only non-time-varying source in the scene. (1/10 sec)

Finally, the profusion of colors in Christmas decorations, and the attractive dashes left by LEDs in photographic traces can provide a rich opportunity for art photography. I provided some examples in the section “The art of photographing time-varying light sources” (page 38) in the chapter on photography earlier in this book. However, they are so pretty that I can’t resist adding a few more samples here.

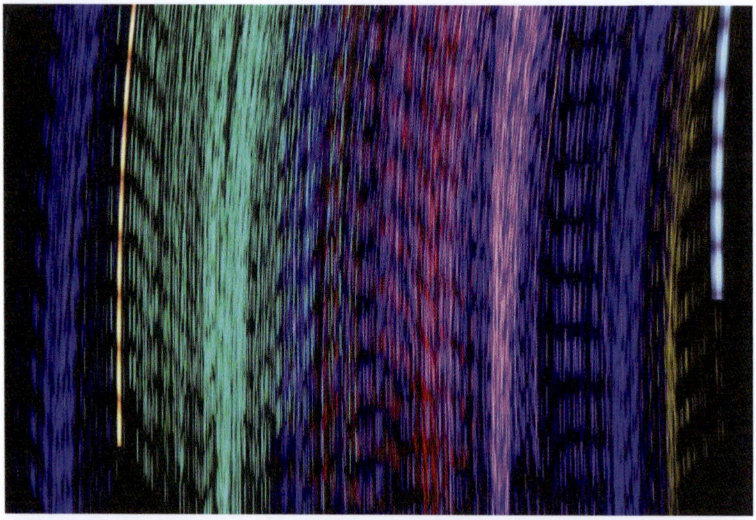

Figure 255: A trace of LED-lit trees in a Christmas display. It looks like two street-lights have made it into the image too. Can you identify their types? (1/10 sec)

Figure 256: Another trace, with some rotation and shaking of the camera during the exposure. (1/4 sec)

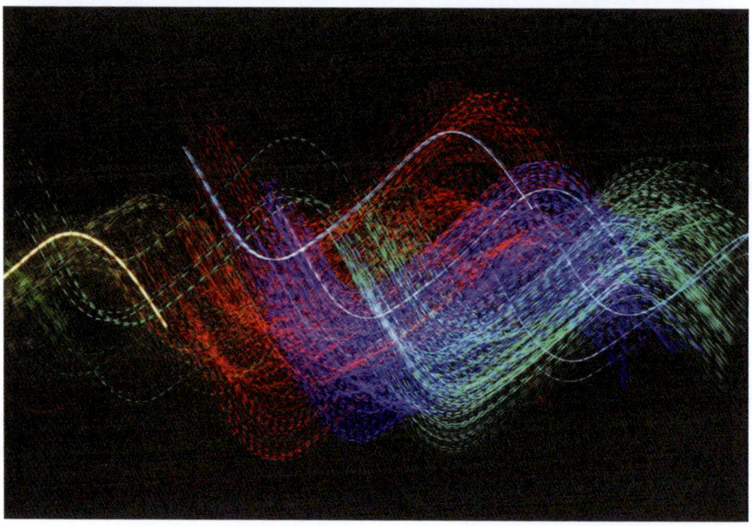

Figure 257: Another trace. In this case, I swept the camera left-to-right, while adding a wave-like motion to it. (3/10 sec)

Figure 258: A downward-swept trace of blue LEDs in a tree. The purple highlights are a result of the behavior of my camera and were not in the original scene. (3/10 sec)

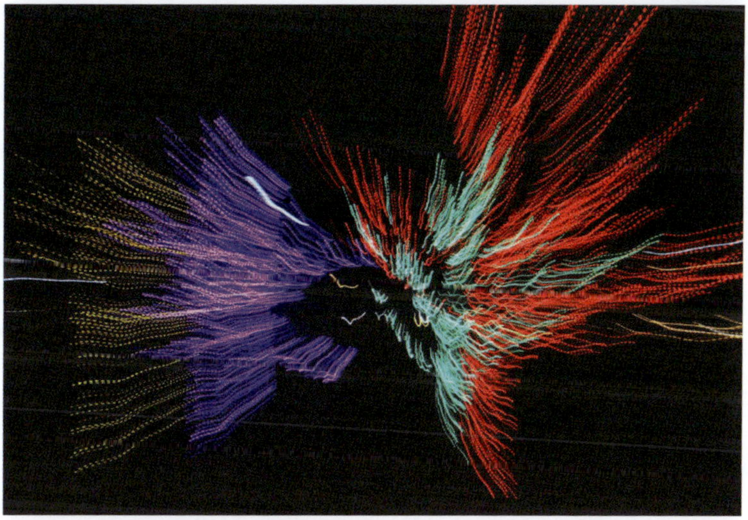

Figure 259: For this shot, I zoomed the lens during the exposure. (1/4 sec)

Figure 260: This tree had orange LED Christmas lights on it. To take this photo, I zoomed the lens and rotated the camera while taking the shot. Although this technique sounds hard to do, it's actually very easy: simply hold the zoom ring steady while rotating the camera! (1/10 sec)

LED stage lighting

Most theater stage lights are incandescent. They are bright and easy to control with a dimmer. They are also not time-varying, so they don't feature in this book.

In recent years, LED-based lights have come on the market and are becoming common. These aren't so bright (though this is changing), and therefore are used either for subtle "wash" lighting (adding a hint of color) or as accent lights in their own right, where the audience actually sees the lights themselves. The main advantage of LED lighting is that with red, green and blue LEDs mounted in one unit the color of the light can be controlled almost infinitely. The human eye responds to mixtures of red, green and blue light by interpreting them as certain colors. For example, red light and green light together are perceived as yellow. Add some blue, and you'll see pale yellow. This is a consequence of the fact that the retina has light-sensitive cells (the "cones") tuned to three color bands: red, green and blue.

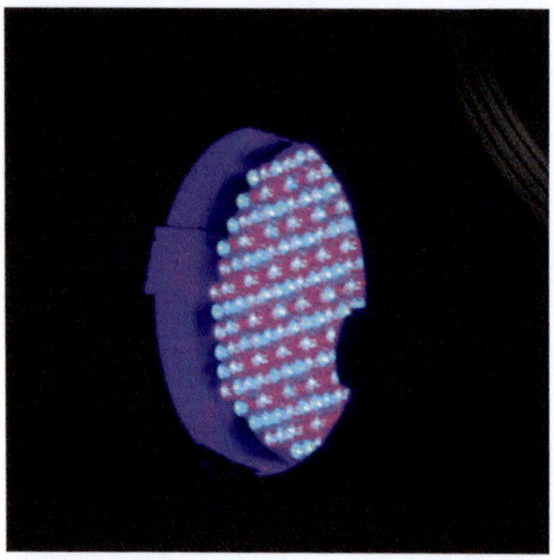

Figure 261: A typical LED stage light, showing an array of individual LEDs, which are controlled electronically to deliver almost any color in aggregate. The camera has not captured the LED colors well; there are red, green and blue LEDs present.

This all means that the apparent color of an LED stage light can be controlled simply by changing the relative intensities of the red LEDs, the green LEDs and the blue LEDs.

However, as you will have learned from other sections in this book, it's not really possible to control the brightness of an LED by varying the drive current. LEDs are, in general, either on or off. Therefore, an electronic circuit is used to flash the LEDs on and off very fast ("modulation"), and the perceived brightness of the LEDs is controlled by varying the "on" time relative to the "off"time. For example, for a red LED to be mid-bright, the "on" time will be the same as the "off" time. To make the LED very dim, say 10% of maximum brightness, the LED's "on" time will be 10% of the total time; the "off" time will be nine times as long.

What makes this type of lighting so interesting for the vision-jiggler is that it's easy to see how the LEDs are strobed. The LEDs are usually arranged in a pattern, with red, green and blue LEDs in regular places. Furthermore, it's common for the different colors of LEDs to be strobed at different times. What this means is that you might see, for example, a series of red dashes with small blue "flowers" in the trace of a light that's shining with a magenta color.

- Next time you can see the LEDs in a colored stage light, jiggle your vision so you can see their time-varying behavior. Look especially for red, green and blue LEDs, irrespective of the perceived color that the light is shining with.

- Look for patterns in the colors which result from the arrangement of the colored LEDs in the stage light. Also look for the relative lengths and timings of the red, green and blue LEDs.

- Observe again when the light is shining with a different perceived color. Even when the light is white, you should see red, green and blue patterns. As the overall color changes, the relative dash lengths for the patterns will change.

Note that if the light is too bright, your retina will be overwhelmed and won't deliver a clear trace. Observe from further away, or turn your attention to a more distant light (see page 21).

Figure 262: A performer on a stage in Las Vegas' Fremont Street Experience. We are interested here in the row of eight LED stage lights above her.

Figure 263: A vertically swept photograph of the eight lights and the performer. It's very clear that the stage lights are time-varying, and that there's a dashed pattern to the traces. By chance, the camera's sweep ended just before the shutter closed, which produced the un-swept image of the performer, producing this unusual and evocative image. (1/10 sec)

Fireworks

Although fireworks make a captivating visual spectacle, they are not particularly rewarding to the observer of time-varying light sources. The reason for this is that almost all the sources of light in a fireworks show are too long-lived, which means that traces will not deliver any additional information. One might as well throw illuminated light-bulbs in the air (see page 69 for why this wouldn't be too interesting). (Actually, throwing light-bulbs in the air could be *very* interesting, but not for optical reasons.)

To somewhat retract what I said above, there is a lot of time-varying activity in fireworks, but the combination of complexity, detail, nearby light sources and short duration render it almost impossible to make any useful or rewarding observations by eyeball jiggling. That's best left to photography.

There are some rare exceptions, so it's worth jiggling occasionally when observing fireworks. For example, some fireworks sparkle with extremely short-lived white or silver points of light. These become short white trails when jiggling.

Figure 264: A long exposure photograph of a firework. The amount of detail in the silver and golden hair-like regions is amazing , and is a testament to both spacial and temporal complexity. (1.6 sec)

Figure 265: There are at least two fireworks recorded in this image. The upper left firework shows an unusual dashed pattern which indicates that the outermost lights were flashing periodically. Since these are driven by combustion, it's unlikely that their light was turning on and off. I suspect that in fact they were shining only on one side of each projectile, and were spinning. (1.6 sec)

Laser light shows

In 2009, we had the opportunity to visit the Crazy Horse Memorial in the Black Hills of South Dakota. At the end of each day (in the tourist season, at least), as dusk deepens, a sound and light show is performed on the mountain, which tells the story of Crazy Horse and the history of the conflicts and relationships between Native Americans and white settlers.

Among the types of lighting used in the light show is a laser display, which "draws" images on the side of the mountain. It's very effective, and a good way to show colored images which tell the story.

Figure 266: A moment from the light-show at the Crazy Horse Monument in South Dakota. The image is "drawn" on the rock of the monument itself, to give an impression of what the monument will look like when complete. Photograph used with permission.

The images have the appearance of line drawings. The reason for this is that the laser beams are deflected by mirrors which are positioned under the control of a computer. To project an image, the mirrors are tilted rapidly through a sequence of angles which cause the laser beam to trace out the picture on the mountainside. In order for the picture to persist for the observer, the mirror must retrace the outline many times per second, to take advantage of the retina's persistence of vision. Guess what? The image on the mountain is a time-varying light source!

When vision-jiggling at a laser-drawn image, two things should be apparent:

- Firstly, it should be clear that the image is being redrawn several times a second. A broad sweep of the eyes should show multiple images of the drawing.
- Secondly, smaller jiggling movements of the eyes or mirror can result in the drawing appearing distorted. This is because this technology relies on the retina remaining in one place while the image is painted by the moving dot of light (this is similar to CRT televisions, discussed on page 81). If the retina moves during the time that the image is traversed by the laser, the resulting drawing will appear distorted.

Las Vegas: a portrait

Few places on planet Earth display such a wild exhibition of exotic lighting as Las Vegas, Nevada. The big casinos compete for visitors' attention with eye-catching signage and impressive light displays. Cutting-edge lighting technology – almost all of it dynamically changing – is deployed across the sides of whole buildings, making the Strip a rewarding destination for the observer and photographer of time-varying light sources. Since so many amazing sights are gathered in one location, I have chosen to present some of them in this special section of the book.

I shall start in Paris – or at least, at the casino of that name – with a pair of photographs of the reduced-scale Eiffel Tower. The lighting itself is relatively mundane: pressurized sodium lighting (see page 50); but the effect is striking.

Figure 267: *(left)* The "Eiffel Tower" outside the Paris Hotel in Las Vegas. It is illuminated with pressurized sodium lights.

Figure 268: *(right)* Although pressurized sodium lights are not particularly interesting, a sideways sweep of the Eiffel Tower produces a striking image. (1/10 sec)

A little south of Paris you'll find Planet Hollywood, a distinctive amoeba-shaped building with dark domes arranged in a regular pattern on its walls. The most eye-catching features are the signage displays – large ones on the ends of the casino, and a long ribbon-shaped one which hugs the building above street level. This latter display is implemented with red, green and blue LEDs. What's impressive is that the whole display illuminates the LEDs 120 times per second, with each "on" period showing a frame of the video content. After the flash, the entire ribbon goes completely dark, though of course without vision jiggling, you would never see this.

Figure 269: A night-time photograph of the Planet Hollywood casino. Note the long ribbon-like display that runs across the front of the building and the large display on the left (below the green "ph" logo).

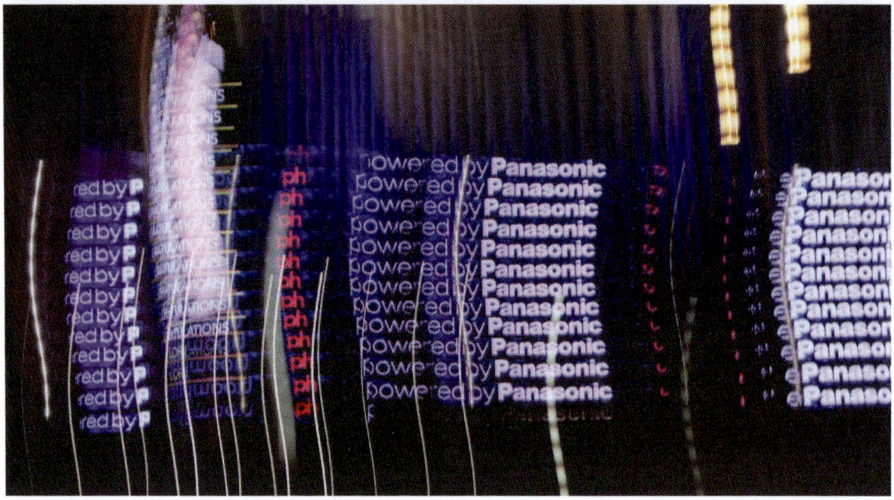

Figure 270: A trace of Planet Hollywood. The long display appears as a sequence of complete images, which shows that all the LEDs in the entire display are illuminated at the same instant. There is no staggered sequence of illumination and no separate red, green and blue phases. Looking at the thin lettering of "powered by", you can see that the letters are *not* stretched vertically, even though the camera was swept vertically. This indicates that the "on" time of the LEDs is really very brief indeed. Although harder to see, the large display on the left also shows a sequence of images, each of which is a complete image of the contents of the display. This display is also flashing all its LEDs simultaneously. (1/10 sec)

Caesars Palace (does anyone know why they use *greek*-style lettering instead of *roman?*) is across the street, a little further north. The large red lettering is illuminated by discharge lighting of some sort, possibly mercury, though the lamps are enclosed in the red letters so it's hard to be sure. These lights aren't in themselves very interesting. What *is* of interest to the observer is that the lights are connected to all three phases of the electrical supply, resulting in the appearance of offsets in the traces.

Figure 271: *(left)* Caesars Palace. For this book, we are interested in the red lettering.

Figure 272: *(right)* Although the letters line up in a row, the dark bands in their traces do not. Close inspection of the traces allows one to determine which letters are connected to each of the three phases. Hint: You might need to place a ruler across the photograph to determine this. (1/10 sec)

Now we come to two LED-based displays of particular novelty. LEDs are extremely versatile due to their ability to be placed in extravagant patterns, and the fact that they can be driven by electronic circuitry or a computer extremely fast to give a wide variety of spatial and color effects.

The first such display shown here is a curtain of LEDs in the Bond bar at the Cosmopolitan. It appears to be a static or slowly-changing pink/white colors. Far from it!

Figure 273: A three dimensional curtain-like array of LEDs in a bar in the Cosmopolitan casino.

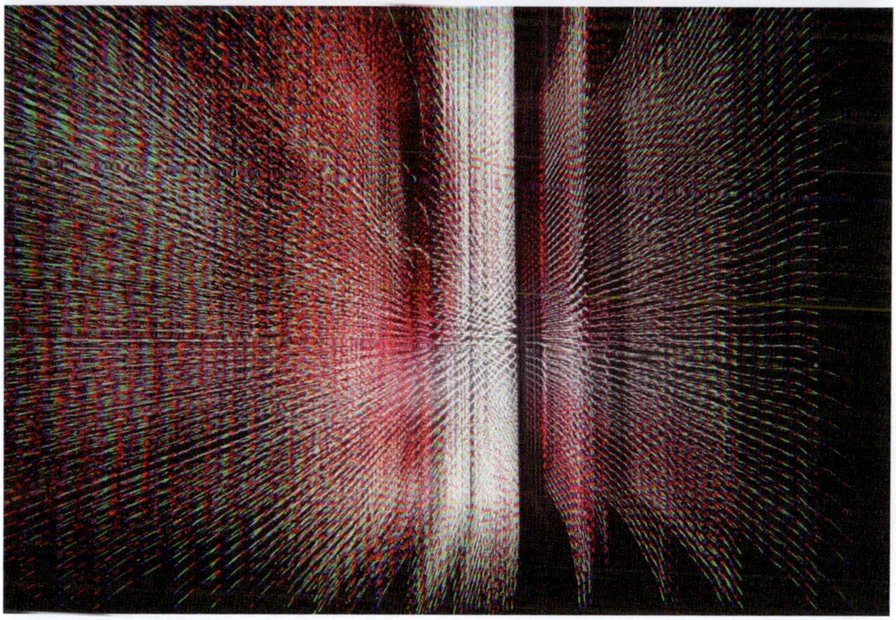

Figure 274: For this image I zoomed the camera's lens while taking the shot. This makes it more of an "art" image than a true documentary one. It's easy to see (e.g. on the right of the image) that although the display is mostly showing shades of pink and white, the LEDs comprising it are in fact red, green and blue, and these are sequenced very rapidly in complex ways to achieve changes in colors and patterns over time. (1/10 sec)

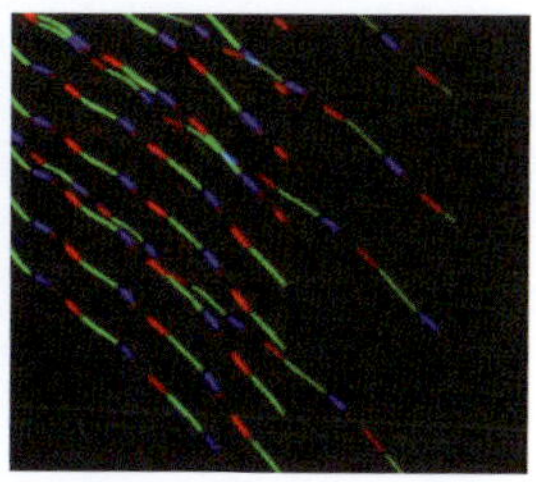

Figure 275: An enlargement of the bottom-right corner of Figure 274, clearly showing the red, green and blue LEDs.

The second LED display in this gallery is on the front walls of the Louis Vuitton store in the new "CityCenter" section of the Strip. White LEDs are arranged in an elaborate pattern and sequenced to deliver eye-catching visual images. These patterns move quite quickly; however, the LEDs themselves are flashing even faster, and that's what a seasoned observer can enjoy.

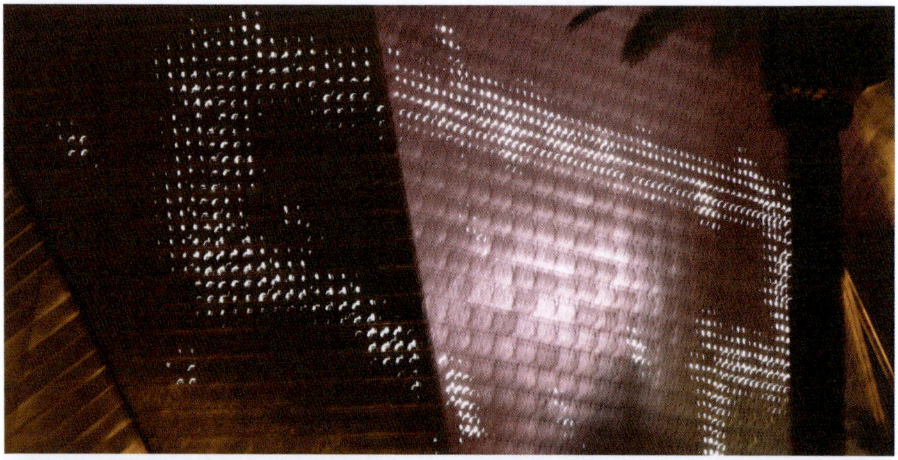

Figure 276: The front of the Louis Vuitton store. The patterns move rapidly across the walls, which makes it hard to do a still "reference shot" and a swept shot of the same pattern. This photograph gives you an idea of a typical pattern. The camera moved slightly ("camera shake"), which is why the LEDs appear as small trails, as I was not using a tripod.

Figure 277: A photographic trace taken by sweeping downwards. It's easy to see that all the LEDs are flashing rapidly. Notice how the pattern of light and dark dashes varies from LED to LED and often within the trace of any one LED. This is due to the computer varying the apparent brightness of the LEDs to contribute subtlety to the pattern. See page 110 for more information on this technique. I have counted the number of dots in a typical trace; there are 30, which means that the LEDs are flashed 300 times per second, seeing as this photograph was a 1/10 sec exposure.

Returning to the Cosmopolitan, this time to the outside of the building, we see that there are many vertical white illuminated strips which accentuate the architecture. These are illuminated with white LEDs, which are time-varying, though you wouldn't perceive this without vision jiggling.

Figure 278: A diagonally-swept trace (though keeping the camera vertical) of the Cosmopolitan. The white strips are obviously time-varying, and are in fact flashing 100 times per second. (1/10 sec)

Figure 279: This image was produced by slightly rotating the camera during the shot. The white strips become double-ended fans, giving an attractive flair to the picture. The viewpoint differs from that of the previous figure. (1/10 sec)

The Venetian hotel has signs on it which say "VENETIAN". The letters are a slightly yellowish white, and appear to sparkle. Although this looks a lot like an old-fashioned sign comprised of small incandescent light bulbs, don't be fooled: this is modern LED technology. How can one tell? By jiggling…

Figure 280: *(left)* Signage on the Venetian hotel. To the eye, it looks like it's comprised of small off-white light bulbs, shimmering slightly.

Figure 281: *(right)* However, the lights are in fact LEDs cycling through a complex sequence: green - red - green - blue - green - red - green. The fact that blue appears less often matches the perception of yellowish-white, since red and green together make yellow, and the little blue that there is shifts the yellow towards white. Note the interesting fact the letters are controlled at different times: the traces do not line up across the sign. (1/10 sec)

Figure 282: Another trace of the Venetian sign, along with other time-varying light sources that paint an almost science-fiction-like scene. Image © 2012 Jessica S. Burton; used by permission.

We can't tour Las Vegas without going into a casino (though I personally like to get out of them as soon as possible as I don't gamble and I don't like cigarette smoke). Slot machines have lights all over them, so it's reasonable to expect that at least some of them are time-varying. This turns out to be true. The following sequence of images was taken from a single viewpoint and moves rapidly from documentation to art.

Figure 283: A typical casino scene. The nearest slot machine has vertical LED-lit bars on each side. These are actually yellow, though in this picture the camera has lost that color.

Figure 284: A documentary picture showing how the vertical bars are time-varying. The perceived yellow color is actually a rapid fluctuation between red and greenish-yellow. Note that there are blue LEDs here too, but they're not used at this point, which is why there's a dark gap in the pattern. (The blue bands that you can see are actually from the neighboring slot machine.) Each time the gambler wins, the colors change and dynamic patterns flow along the bars. (1/10 sec)

Figure 285: One of my favorite pictures of Las Vegas, taken by slightly rotating the camera when shooting. The gambler is still visible, but the world appears centered around him, suggestive of the single-mindedness of the gambling experience. The time-varying nature of the vertical bars results in the attractive fans. (1/10 sec)

Figure 286: Another image of the same scene, taken by a friend whom I was training in the photography of time-varying lights. Image © 2012 Jessica S. Burton; used by permission. (1/10 sec)

Figure 287: Taking this further, we get this amazing visual treat. Rather than rotating the camera around the axis of its lens (as in the previous figure), I swept it in an arc centered on my chest. The exposure is half a second, which gives much more time to paint the traces; notice that there are many more bands in each fan than in the previous photos. (1/2 sec)

Figure 288: Finally, this image was taken by simultaneously rotating the camera and zooming the lens. This sounds like a coordination nightmare, but it's actually very easy: hold the zoom ring still with one hand while rotating the camera with the other. (1/5 sec)

On the corner of Las Vegas Boulevard and Tropicana Avenue stands the MGM Grand, bathed in green light. In front of it stands a large golden lion, behind which is a wall covered with clusters of LEDs which change color frequently. Naturally, to cover a wide range of colors, these clusters contain red, green and blue LEDs, and all of them are controlled by a computer. Vision jiggling reveals some interesting and attractive activity. Here are two images that illustrate this:

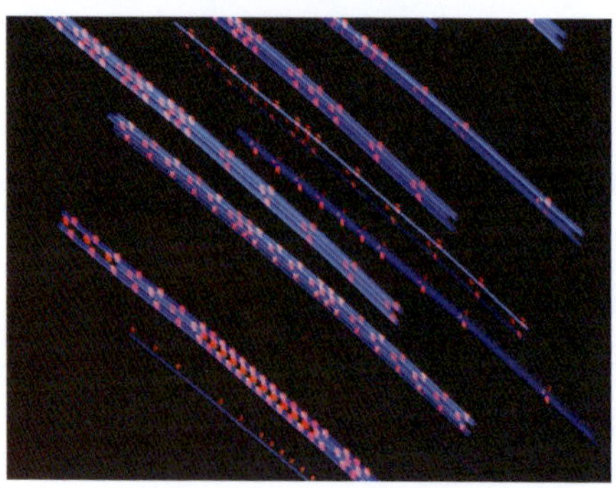

Figure 289: *(left)* The apparent color of the LED clusters was a range of violets. This was achieved by turning on the blue LEDs and adding pulses of red. Notice that the patterns change along each trace: the colors were not constant but were changing continuously. (1/15 sec)

Figure 290: *(right)* This delightful image was taken when the LED clusters were changing apparent colors very quickly, for a twinkling effect. As you would expect, the sequence of pulses is complex. (1/15 sec)

I need to confess that Hard Rock Cafe – just north of the MGM Grand – is one of my favorite places on Earth. Let me qualify that: The *outside* of Hard Rock Cafe is one of my favorite places on Earth! Somewhat like the Cosmopolitan, Hard Rock Cafe has vertical lighting strips at intervals along the building. However, these are anything but white; they are multi-colored and changing extravagantly all the time. To achieve this, the LEDs in the light-strips are sequenced at very high frequencies by a sophisticated combination of software and computer hardware. Now, for the non-jiggler, these vertical bars are quite cool, and may merit a few seconds' attention. When I and a friend discovered them, we were enthralled for twenty minutes, eyeball-jiggling, mirror-jiggling and, of course, taking photographs. Enjoy the following sequence of images, courtesy of Hard Rock Cafe.

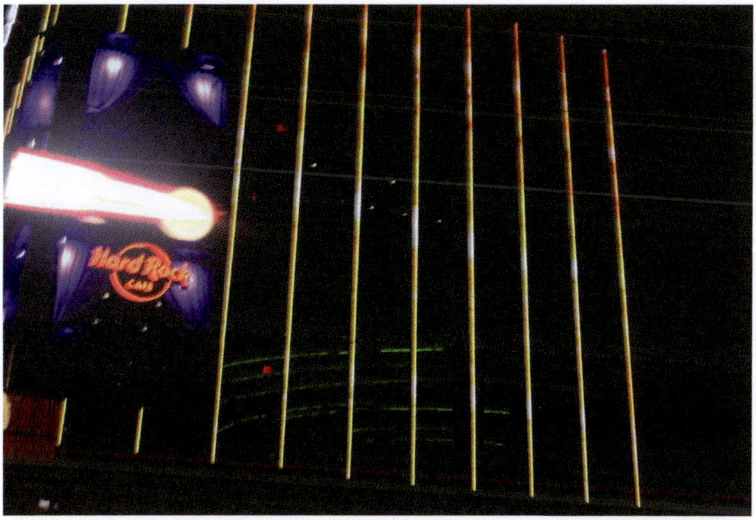

Figure 291: The south side of the Hard Rock Cafe building, showing the vertical illuminated strips.

Figure 292: A blue and purple pattern, stretched sideways into a trace by sweeping the camera sideways slightly during the shot. (1/10 sec)

Figure 293: A yellow-and-black pattern. The diagonal stair-stepping that you can see in the traces corresponds to the black and yellow regions moving up or down the strips. (1/10 sec)

Figure 294: A mostly-green pattern. Note the curious blend regions where the strips have black sections. (1/15 sec)

Figure 295: Many colors, and they're changing. See how there are step-changes in the coloration as you look across each trace. (1/15 sec)

Figure 296: For this shot, I rotated the camera. (1/10 sec)

A new feature of the Las Vegas skyline is High Roller, the giant observation Ferris wheel. At night it's illuminated with changing colors. The source for these is LED lighting, and you would expect this to be time-varying. You would be correct.

Figure 297: A trace of High Roller illustrating the complex sequencing of the LED -based illumination. The white "Sands" logo at the bottom of the picture is comprised of white LEDs, which are also time-varying. You probably won't get to see this view: it was taken from the ballroom loading dock at the Wynn. (1/10 sec)

Las Vegas represents one of the best locations for this type of photography. Aim to spend a leisurely evening after dark walking along the Strip with your camera. And you won't need a tripod!

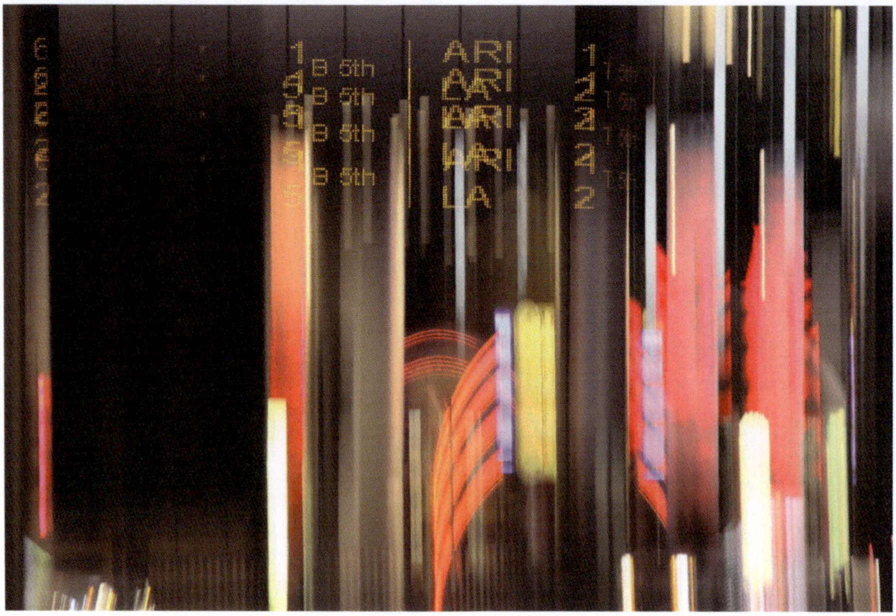

Figure 298: A vertically-swept photo of a scene on the Strip. It's amazing to think that an image this complex can be created just by moving the camera when pushing the button. (1/30 sec)

Natural time-varying light sources

Moths and flies

If you recall from the opening sections of this book, one of the easiest ways to observe a time-varying light source is to have someone wave it around, because then you don't need to jiggle your eyes or a mirror. Thankfully, nature has provided a wonderful light source in the form of moths and flies, which move quickly enough for their images to travel rapidly across your retina.

Of course, moths and flies don't generate light, so you may be wondering why I've included them. The answer is that they flap their wings very fast; and if they are flying in the light of an outdoor lamp or vehicle headlights then it's possible to see rapid variations in the traces they leave across your vision.

A good way to observe moths is to drive on a windless summer night along a country road. Any moths caught in the beam of your headlights will be moving very fast, because of the combined motion of the moths' flight and your car. Photography, however, is easier to achieve when watching moths flying around an outdoor lamp, especially if you can keep the lamp itself out of the camera's shot.

Flies also flap their wings, of course, but are usually smaller, and their wings may flap so quickly that it's hard to see the variation. If you are into photography, you can actually measure how fast they flap their wings: as long as the fly is in the camera's shot from the start to the end of its trace, you can count the flaps and correlate it to the exposure time (shutter speed) at which the photograph was taken. For example, if your photograph's exposure time was 0.1 seconds, and the fly's trace has 35 flaps, it's flapping at 350 times per second! This may sound quick, but think of the sound of a gnat or mosquito when it's

Figure 299: This photograph records the flight paths of two moths: a larger one on the left and a smaller one that's in the middle of the image; their paths converge at the top of the photograph. Note that there was no need to move the camera in this shot: the motion of the moths was sufficient to paint these traces. The lamp is hidden behind the corner of the brick wall on the right. There are approximately 107 wing flaps in the trace of the smaller moth, and since the trace is entirely contained within the image, we can determine how fast it was flapping: just over 350 flaps per second! The fuzzy white patch by the wall is another moth flying in a tight path below the lamp. Below this, you can see thin trails left by very small flies. This image doesn't have enough resolution to allow us to discern these flies' wing flaps. The branches of a pine tree can be seen in the background. (3/10 sec)

flying: it's a high-pitched whine, which must be at several hundred hertz or even a kilohertz... which would be a thousand flaps per second!

Figure 300: Peter Steward's photograph of moths in front of a lamp in Tsavo West National Park in Kenya. This photograph (shown somewhat cropped here for detail) won the "ecology and society" category of the British Ecological Society's photographic competition of 2014. Photograph © 2013 Peter Steward. Reproduced by permission. (2 sec)

Lightning

Lightning's most interesting time-varying behavior arises from the fact that most bolts of lightning actually consist of repeated strokes along the same path. This is due to the fact that the initial stroke creates an ionized channel in the air, and subsequent strokes discharge along this channel which serves somewhat as an electrical conductor.

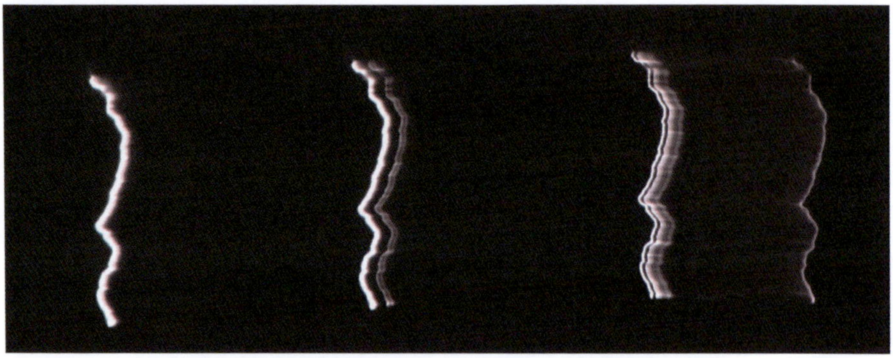

Figure 301: A cloud-to-ground lightning strike, photographed while the camera was swept sideways. The exposure time was 1/4 second. Note the multiple strokes which all follow the same path. The apparent stroke on the far right is actually a photographic chimaera: there's a wide area which records a sustained period of dim glowing light; the right edge of this glow is actually where the camera stopped moving as it reached the end of its horizontal sweep, allowing the dim glow to form a localized, brighter image. Close inspection of the glow area shows that the camera actually headed back in the opposite direction for a short time.

However, observing this behavior is not easy. The main problem is that it's impossible to predict when a suitable bolt will occur. This means that naked-eye observation requires sustained and continuous jiggling – ideally in the direction that's perpendicular to the bolt itself, in order to separate out the images of the individual strokes. In my case, I am much better at jiggling up and down, so to achieve a sideways movement of my eyes, which is required for observing cloud-to-ground lightning, I had to lay my head on its side. This was disorienting and hard to maintain.

Additionally, because lightning is unpredictable and visually arresting, it's easy to become like a deer in the headlights and to stop jiggling at the moment of the stroke. This defeats the object of the observation; and so one must train oneself to keep jiggling no matter what.

What is one looking for? Well, a correctly made observation consists of seeing the same shaped stroke of lightning apparently in multiple places of your field of view. This shows that multiple identical strokes occurred while your retina was moving.

- Try using a mirror. As long as you have the hand or arm coordination to keep up a regular oscillation, this may well prove easier than eyeball jiggling.

- Although cloud-to-ground lightning is the most rewarding, mainly due to its brightness and relatively long duration, cloud-to-cloud lightning can also show multiple strokes along one path. For vertical eyeball jigglers like myself, this tends to be easier to watch for.

Photographing multiple strikes is also not easy. I have used two approaches. Figure 301 is a still photograph taken by repeatedly clicking the shutter button while sweeping the camera left to right. I did this approximately twice a second for up to a minute at a time, like a machine. The idea was that in some shots there would be a flash of lightning, and that turned out to be the case: out of about 320 shots, I captured 12 photographs that were potentially usable!

A second approach, shown in Figure 302, is to use a video camera, or a digital SLR that can shoot video. Video is, after all, a way of capturing 25, 30 or even 60 "photographs" per second for an extended time. Start recording, and flap the camera up and down (or side to side) in a rapid, regular motion which should be kept up during several lightning events.

You will not know what you have captured until you look at the video file in software which allows you to step through frame by frame. Once you find a good frame, it may be a good idea to export it as an image which can be optimized (say, for contrast) in an application such as Adobe Photoshop or The GIMP. The quality and pixel resolution may not be as high as a still photograph, but the results can be impressive.

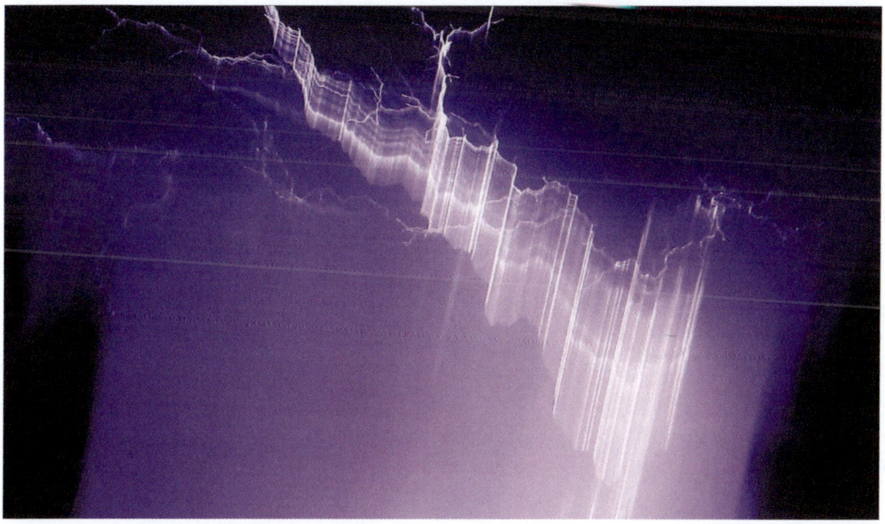

Figure 302: An image taken from a video that was recorded while the camera (a Canon EOS 70D) was regularly and rapidly swept upwards and downwards for over two minutes. This frame, which shows complex and multi-stroke behavior in cloud-to-cloud lightning, was processed using Photoshop to enhance the contrast. Since the video was shot at 30 frames per second, the exposure time was 1/30th of a second or shorter.

Finally, for both still or video approaches, focusing the camera may be difficult due to the fact that when there's no lightning, there's nothing to focus on. Turning off auto-focus and using trial and error may be necessary.

Stars

Twinkle, twinkle little star could possibly be rewritten as "Behave in a time-varying way little star". It turns out to be true: vision jiggling at the night sky does indeed show the twinkles to be random variations in brightness.

The stars themselves are not twinkling: observed outside of our atmosphere, they appear completely steady. Twinkling – or more correctly, scintillation – is a result of perturbations in the atmosphere causing the star's light to spread and bunch. It's a little like looking at the bottom of a sandy stream or a swimming pool in sunlight: the ripples on the surface cause patterns of lighter and darker illumination on the bottom (these are called "caustics").

To observe twinkling stars, you need three things: a cloudless night, a dark location and dark-adjusted eyes. To find a dark location in the United States, Southern Canada and Mexico, this website is very useful: http://www.jshine.net/astronomy/dark_sky/

The easiest stars to observe are – of course – the brightest ones, so it pays to know how to find, say, Sirius. There are plenty of maps of the night sky on the internet which can help you to find the brightest stars; however, simply going outside at a dark location will work too. Those of you with smartphones will be able to use Google Sky Map to identify the stars you are looking at. Scintillation is stronger for stars that are lower in the sky than for the ones that are close to the zenith. This is simply due to the "amount of atmosphere" that the light has to travel through before it reaches your eyes.

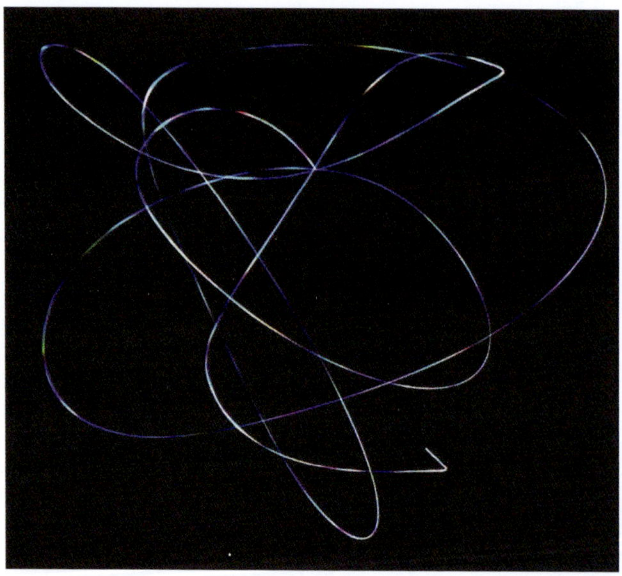

Figure 303: A photographic trace of Sirius, the brightest star in the night sky. Rather than using a simple downward sweep, I chose to move the camera in a series of smooth loops. The colors are very obvious in this image (see the text). The naked eye cannot see these colors. The *twinkling* that the naked eye *can* see is represented in this trace as the most obviously brighter and dimmer regions. This photograph has been processed in Photoshop to enhance the contrast, which helps to ensure it is printed clearly. (1.6 sec)

Vision jiggling will reveal the largest variations in brightness that constitute the scintillation that you can see. The retinal trace will appear as a white line with random brighter and darker regions. However, in reality, the variation is much faster than you can detect, and it's *colored!* The "ripples" in the atmosphere function a little like prisms, causing an effect called "chromatic dispersion", which separates the colors of white light. The narrowness of the star's image means that the colors don't often overlap (which would make white), giving rise to a rapid, psychedelic dance of color.

Sadly, however, you can't usually see these colors with the naked eye, for two reasons:

1. There is not enough light from the star, which means that the cone cells in your retina (which are responsible for detecting colors) are not stimulated; only the rods are stimulated, and rods do not detect color.

2. The colors change so swiftly, about once every 0.3 to 10 milliseconds, that even if they were brighter, the eye would blend them together to white. You might think that eyeball- or mirror-jiggling would separate the colors out along the trace, but then reason #1 comes into play and you can't see them.

However, if you take a telescope or an SLR camera with a long telephoto lens (200mm or higher) and look through it at the star, the intensity is increased so there is now enough light to stimulate your cones. Jiggle the telescope or camera and you may be able to see the colors.

Of course, a camera doesn't have the "cones *vs.* rods" problem, and can capture the colors just fine. And what colors they are!

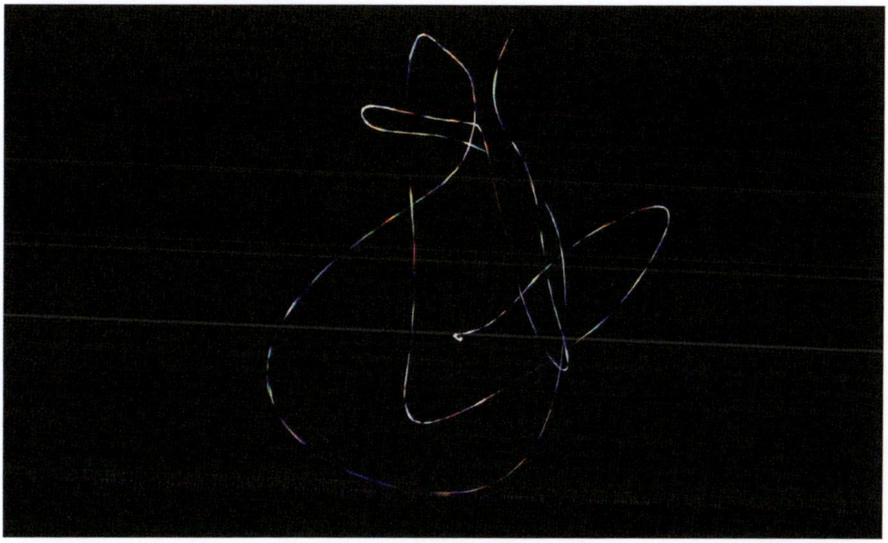

Figure 304: A photographic trace of Arcturus, which was lower in the sky than Sirius at the time I captured these images. Less bright than Sirius, it leaves a fainter trace; however, the colors are, if anything, even more pronounced. (1.6 sec)

Planets

The easiest way to be sure that you are looking at a planet and not a star is that stars twinkle and planets (usually) don't. Vision jiggling confirms this, with no detectable time-varying behavior. The main optical difference between stars and planets is that the naked-eye planets in the solar system have *width*, i.e. they appear as discs rather than points. This means that the images are large enough that the atmospheric "ripples" do not distort or color them. Stars, on the other hand, are far enough away that they essentially have *no* width when seen from earth, and are susceptible to scintillation.

Figure 305: A photographic trace of Venus. It's clear that there is essentially no time-varying behavior, as the trace is an almost constant, unbroken white. (1.6 sec)

Pulsars (if only!)

Probably the grandest time-varying light sources in the universe are, sadly, out of our reach, and the vision-jiggler can only imagine what they would look like. These are pulsars, which are nature's own celestial strobe-lights.

Pulsars are collapsed stars, which have a number of interesting properties:

- They are super-dense (composed entirely of neutrons) and have very strong magnetic fields.
- They emit radiation, usually including visible light.
- They are spinning, in some cases extremely fast – up to *700 times per second!*
- The combination of high rotation speeds and strong magnetic fields causes the radiation to be emitted in two beams, somewhat like a double-ended lighthouse – or maybe it's easier to imagine one of those rotating lawn sprinklers.
- When Earth happens to be in a location illuminated by one of the beams as it sweeps past, the pulsar appears to be flashing at a rate determined by its rotation. (This gives rise to the name "pulsar": it's a star that appears to pulse.) There is a pulsar, for example, in the center of the Crab Nebula, which appears to flash 30 times per second.

Figure 306: A composite image of the Crab Nebula showing the X-ray (blue), and optical (red) images superimposed. The Crab Pulsar is the bright star in the center of the circular nebula. *Source: NASA & ESA*

I have always wanted the chance to jiggle at a pulsar (*that* would look interesting on my grave stone!). Unfortunately, there are no pulsars anywhere close enough to the earth to be observable by the human eye. Even if we could see them, they would be so faint as to leave no trace on the retina.

I suspect too, that even were we to see a pulsar through a large telescope, it would prove impossible to jiggle at it and see a trace. Modern telescopes operate by integrating images over time (i.e. they accumulate photons in a light detector); and such rapid time-varying behavior would be obscured.

The best we can do is to simulate what a pulsar might look like were we in a space-ship and close enough to see it alone from other stars.

To achieve this, we can refer to the published intensity plot, shown in Figure 307. Then, what's needed is a software tool which allows a dynamic brush to be configured so that the brightness along the path changes approximately in accordance with the intensity plot. Another complication is that most software applications that support dynamic brushes, such as Adobe Photoshop and The GIMP, do so by *distance traveled*. This is not appropriate for emulating a trace, because the brightness variation needs to be time-based, not distance based. In other words, the dashes and gaps need to become longer if the drawing tool is moving faster. Thankfully, the free software MyPaint supports this mode of painting, which allowed me to create the synthetic trace shown in Figure 308.

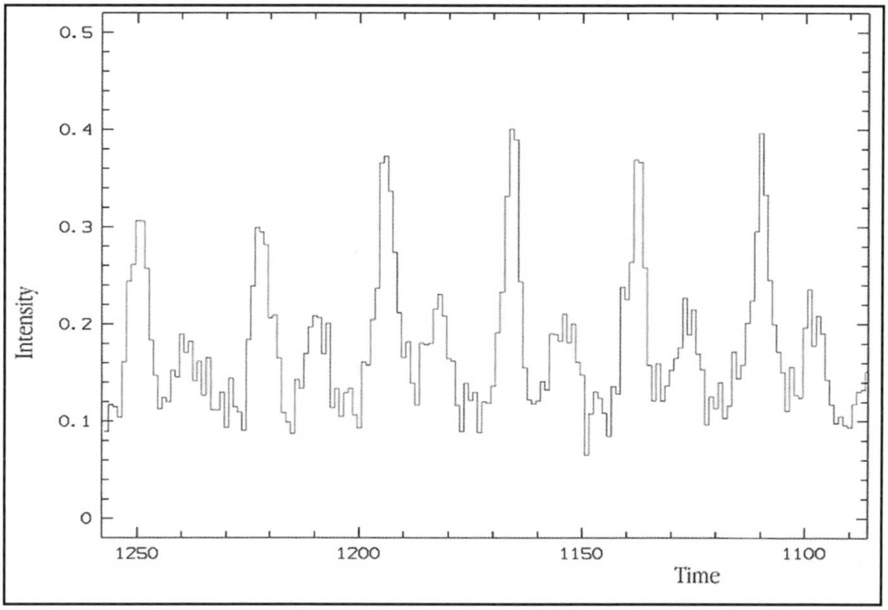

Figure 307: A plot of the intensity of the Crab Pulsar over time (in milliseconds) as seen from Earth. A total of six cycles are shown here. Each cycle has a bright peak and a dimmer peak. The stepped appearance is a result of the times at which the intensity was measured (sampled) and is not a true feature of the actual intensity. Source: http://www.eso.org/public/archives/images/screen/eso9948i.jpg

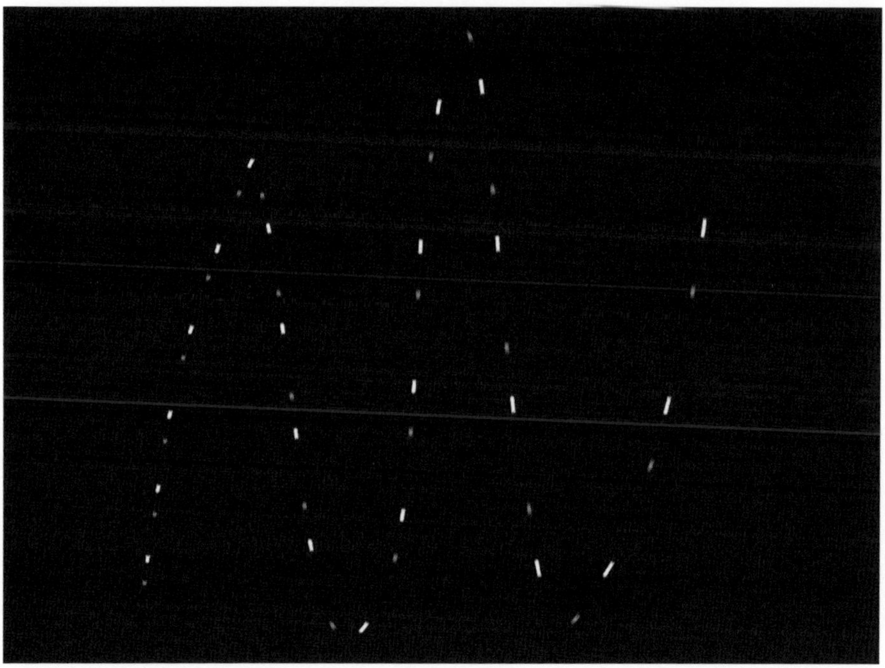

Figure 308: A synthetic trace representing what one might see if able to sweep a camera at the Crab Pulsar, where the direction of view is the same as that from Earth. Created using MyPaint.

Time-varying reflections and interruptions

Unlike the rest of this book, the light-sources in this section are not time-varying. However, these observations are still interesting for the vision-jiggler: they arise because of a high-speed variation in something that's reflecting or interrupting a steady light source.

Water drops in sunlight

I remember standing one day on a platform of the railway station in Weybridge, England. It had been raining, but now the rain had stopped and the sun had come out. While waiting for the train, I looked across at the platform on the other side of the tracks, and I idly noticed a drop of water as it fell from the canopy above. What caught my attention was the fact that the drop appeared as a series of bright dashes, and since I'm a seasoned vision-jiggler, I noticed it immediately. (This, by the way, is what you should aspire to: the ability to see interesting time-varying behaviors even when you're not looking for them!)

Figure 309: Weybridge Station (on a dry day).

"Oh," I thought, "it's being illuminated by a time-varying light source." But then it struck me: the light source was the sun; and the sun's not time-varying! I think we would know by now if it were a pulsar (see page 144), and it's certainly not one.

So if the light source wasn't time-varying, how come the water drop appeared as a series of dashes?

The answer is actually very simple: the drop was not spherical, and was tumbling (rotating) as it fell. This means it would sometimes catch the light, and sometimes not. The result to my eyes was a time-varying appearance, which I could see and interpret.

When the sun is shining on rain, or where there are water drops falling in sunlight, see if you can see any that show time-varying behavior. Are all time-variant drops the same? Does a drop change its time-varying behavior as it falls? Note: you won't have to jiggle for this observation because the raindrops are falling across your field of view. Just watch them and keep alert for dashed trails.

Lawn sprinklers are a good source of water droplets! However, not all water droplets are spinning, and some sprinklers generate a relatively low proportion of spinning drops. With careful observation in sunlight you can determine if spinning drops (which leave regularly-dashed traces) are common or not.

Figure 310: A photograph of water droplets ejected from a lawn sprinkler. Although the image is somewhat confused by the background, it should be clear that most of the traces show regular dashes, indicating that these drops were rotating and non-spherical. Close inspection of this image shows colors in the droplets, which is a pleasant surprise as this effect is not usually visible to the naked eye due to the droplets moving too quickly. (1/10 sec)

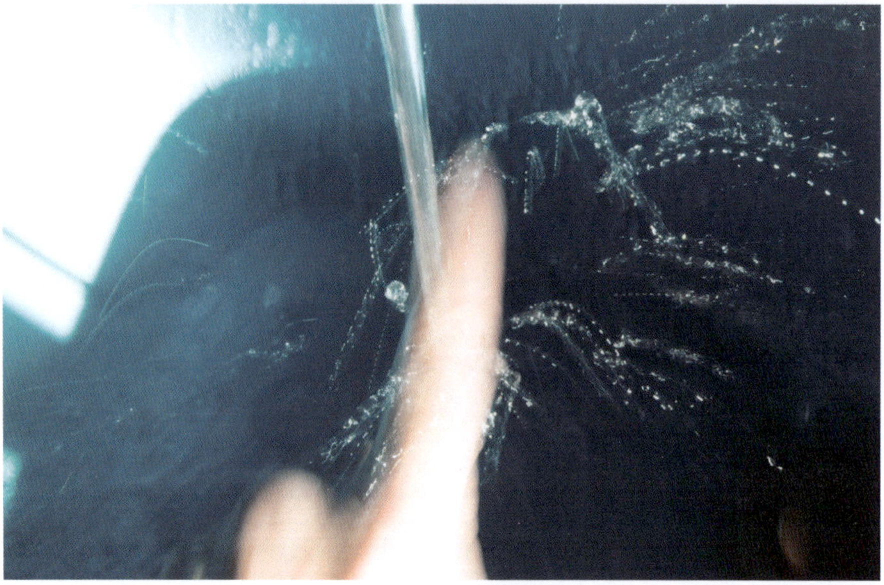

Figure 311: A photograph taken by lifting my finger in the stream of water from the tap in my bathroom sink. The image was illuminated by sunlight. Most of the scattered drops show clear periodic brightness and darkness, indicating that they were (a) spinning and (b) non-spherical. (1/10 sec)

Chrome wheels

Figure 312: A chromed wheel on a 2008 Chrysler 300M.

One source of time-varying reflections is from shiny rotating things that glint in the sunlight. While it might look like the gleaming reflection is constant, it may indeed be flashing on and off rapidly as the object rotates, now catching the sunlight, now not. Two examples of this come to mind: chrome wheels on, say, a truck and rotating vents on the roof of a building.

This type of wheel, which is often seen on

pickup trucks in the USA, is a good source of time-varying reflections. However, it's only observable when the wheel is rotating, which means (usually!) that the truck must be moving along the road. Standing by the roadside and vision-jiggling at the truck should allow you to see the reflections as a series of dashes (actually, each flash will have the shape of the glint as shaped by the contour of the wheel where the sun reflects off it). However, since the truck's moving and you aren't, you won't see the wheel for very long.

You may also get good results if you can train yourself to keep your eyes still while the truck is passing across your vision. This relies on the motion of the truck to keep the image of the wheel moving across your retinas.

Another approach is to observe the truck's wheel from your vehicle as you proceed along the road alongside or near the truck, such as on an interstate or other multi-lane road. Of course, it's a *really* good idea to do this when someone *else* is driving!

Figure 313: A photograph of the 300M while it was driven past the camera in sunlight. The wheel's complex surface gives rise to a sequence of shapes where the chrome catches the sunlight. The periodic nature of the pattern comes from where each spoke catches the light in turn. (1/10 sec)

Hedges and grass seen from a moving car

I live just east of the urban corridor in Northern Colorado, at the western end of the great American plains. Looking west from my house, I can see "the ridge where the West commences"[1], at this latitude, at least. To the East are wide open spaces, mostly farmed, though urban sprawl is taking bites out of the countryside.

One consequence of the not-quite-flat landscape is that it's usual to see street lights or farm lights at a distance of one to several miles, with little in the way to obstruct the view. Actually, that's not *quite* true: most of the rural roads around here have tall grasses and plants growing along the sides; and this brings about an interesting effect which is visible to the seasoned observer of time-varying light sources.

It often happens that when traveling in a car at night along such a road, the grasses and shrubs lie between your eye and the distant street light. Vision-jiggling will reveal two things going on:

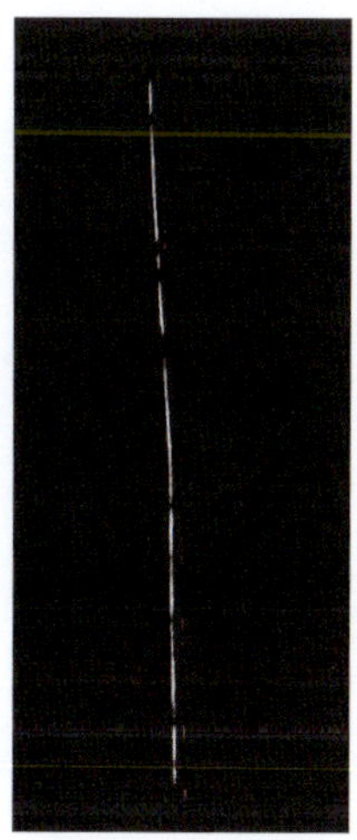

Figure 314: A bright white light captured through the branches of leafless trees, by sweeping the camera while shooting from a moving vehicle. As the car moves, twigs and branches interrupt the view of the light. Sweeping the camera captures the history of these interruptions during 1/10 of a second. Since the branches and twigs have hard edges, the interruptions are mostly abrupt. The primary impression here is that the interruptions are random in nature. Note that this particular light does not show time-varying behavior of its own. It's a powerful lamp on an oil drilling site.

1 "Don't Fence Me In" by Cole Porter and Robert Fletcher

- The street light's own variation (if any) will be apparent (see page 49).
- The grasses and other plants will interrupt the light in a rapid, random way. This produces a striking effect, shown in Figure 314, though it's more interesting seen with the eye: for some reason, photographs don't quite do it justice.

Since the leaves and stems are (visually) hard-edged, so are the random dark features in the trace.

In fact, at night, you often wouldn't know that there are grasses there; but a quick jiggle will reveal their presence to you by showing this effect.

You might want to make this observation when someone else is driving!

Wire fencing seen from a moving car

A variant of the hedges effect described above is when you're driving past a diamond-pattern wire fence. The visual characteristics of such as fence are: the pattern is completely regular, the wires are diagonal, and the wires are sharply defined. This gives rise to a very characteristic pattern of time-varying interruptions which is easy to observe with vision-jiggling. Photography is a little harder to achieve, especially with regard to pre-focusing; but with care and patience it's possible to obtain great images.

Figure 315: A wire fence of the type discussed in this section.

Figure 316: A vertically-swept photograph of the headlights of cars on a busy interstate road, taken from a car that was being driven across a bridge over the interstate. The bridge had diamond-pattern wire fencing along its sides. The diagonal dark lines are very apparent, recording how the wires interrupted the light from the headlights during the swept exposure. (1/10 sec)

Lights seen through a fan or propeller

Any time you can see a light source where there's a fan between it and your eyes, vision jiggling will reveal interesting time-varying behavior. This is true even for sources that are not time-varying, such as incandescent light bulbs. The fan serves to rapidly and repeatedly interrupt your view of the light.

Airplane propellers also work in this way. In 2013, I flew in a Bombardier Dash 8, which is a twin turbo-prop aircraft, in a perfect seating position for looking through the spinning propeller at distant lights. While it's possible to obtain interesting images of the propeller blades during day-time flight, the most interesting eyeball jiggling and swept photography opportunities come when the aircraft is flying low over a city at night, or on the ground, taxiing across the airport, also at night. The interruptions due to the passing propeller blades are very clearly seen, and deliver traces that are quite different to most that you would see.

Figure 317: A Bombardier Dash 8 8-400 of the type mentioned in the text and in the following figure. Note that there are six propeller blades for each engine. (Photographer Laurent Errera.)

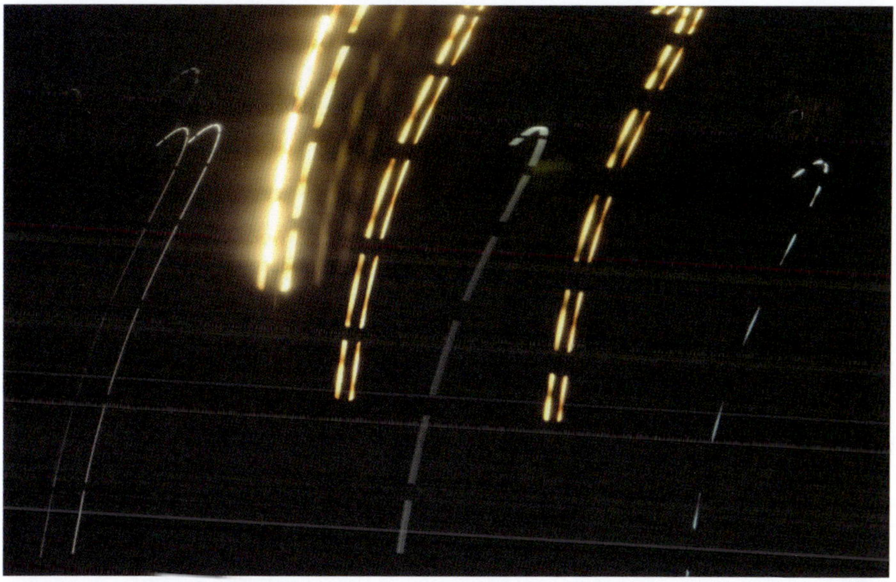

Figure 318: A photographic trace of lights seen through the blades of a propeller on a Dash 8. The photograph was taken from a passenger seat as the airplane landed at night. The brightest traces are from pairs of pressurized sodium street lights (see page 50). As you would expect from these lights, you can see the regular transitions from pale yellow to a darker yellow color. However, these traces are interrupted by regularly-spaced dark bars, which occurred when propeller blades passed in front of the lights. The sharp edges of the interruptions are due to the sharp edges of the blades. We can estimate the rate at which the blades were passing in front of the camera. The dark yellow regions of the traces occur 120 times per second (twice the 60Hz frequency of the AC electrical supply). By careful inspection, we estimate that the black interruptions are occurring a little closer together (therefore faster) than two dark yellow regions. Careful measurement of the distances between the black interruptions and the dark yellow regions confirms that blades were passing in front of the camera 70 times per second. With six blades on the engine, we can deduce that the engine was turning 11.5 times per second, or 700 rpm when taxiing at the time the photo was taken. There are two additional interesting features of this photograph. Firstly, if you look at the size of the black interruptions, you can see that they become larger towards the right side of the photograph. This is due to the shape of the propeller blades, with the thin, white, left-most trace being interrupted by the tip of the blade only. Secondly, it's interesting to note that all three pairs of pressurized sodium lamps show non-aligned dark yellow regions, with the displacement being by one third. This shows that the pairs of lights are wired to different phases of the three-phase electrical supply. See page 51 for a discussion. (1/8 sec)

This effect is not limited to fans and propellers; many rapidly rotating spoked wheels would show the effect, though the spokes of a bicycle wheel are likely to be too narrow to effectively interrupt the light.

Because of the hard edges of fan and propeller blades, the edges of the observed pulses of light are sharply defined. They may also have a curved look to them, partly due to the shape of the blades and partly due to the fact that the blades are rotating during the time that you can observe the light between them.

Another instance where a light source is interrupted by a fan is an office projector, which is described on page 95.

Figure 319: *(left)* A vent in the wooden roof of a building. This vent has a fan in it, which is essentially invisible to the casual observer; in fact, only by noticing the retinal trace did I observe that there was a running fan. The light is sunlight, which is of course not time-varying.

Figure 320: *(right)* A photographic trace of the vent reveals the periodic interruptions of the fan blades. This image was taken from the same viewpoint as the previous figure. Notice how the fact that the blades take time to traverse the opening leads to the light appearing to tilt at a different angle to the original shape of the bright opening. The appearance of this image is very close to what I observed by eyeball jiggling; and the faster I jiggled, the more tilted the bands appeared. These photographs were captured with my cell phone, which is much harder to control than a digital SLR camera. Also, the EXIF data appears not to be attached to the JPEG file, so there's no way to know the exposure time and hence the rate at which the fan blades are interrupting the light.

Conclusion

Thank you for coming with me on this journey into the fourth dimension. As you can probably tell, writing this book has been a very personal project, driven by the desire to give others the opportunity to catch something of my passion. To the best of my knowledge, this book is unique, being almost certainly the only book anywhere that addresses the subject of observing time-varying light sources. It may be a little off the beaten track; but as lighting technology advances there will only be more time-varying light sources around us, not fewer.

I encourage you to head over to the web-site and share your experiences on the forum, or contact me directly. I would love to hear from you if you have found your view of the world enriched by this book. Remember - this Field Guide is not exhaustive – there are other light sources out there to be discovered and enjoyed.

I wish you happy jiggling!

Web-site

The web-site that accompanies this book may be found at **timevaryinglights.com**

It provides videos that describe how to perform eyeball- and mirror-jiggling, how to take swept photographs, and more. There is a forum where you can join the discussion, ask for advice or share your experiences. And you can contact the author directly.

Acknowledgments

Thanks to my daughter Zoë for being a great model.

Thanks to Zach Gillis at Ed Carroll Audi in Fort Collins for standing in the cold while I took photographs of tail lights.

Thanks to Jonathan Turner for letting me photograph his plasma TV.

Thanks to Gary Kegerreis for driving past me multiple times in his 300M.

Thanks to Ricardo Sarti for letting me photograph his GMC Acadia.

Thanks to all those who have provided permission to reproduce their photographs.

Thanks to my family for supporting me in this endeavor.

And thanks to my parents for buying that oil-filled radiator.

Image credits

All photographs were taken by the author, with the following exceptions:

Author portrait on page 3: Photograph © 2012 Hyeon Na Woo. Used by permission.

Figure 2: © Glen Dimplex Group. Used by permission.

Figure 4: Photograph © 2011 Nathaniel C. Liedl. Used by permission.

Figure 12: Images courtesy of PhaseSpace Inc. These were captured at SIGGRAPH 2012 by a high-speed video camera at 450 frames per second; every fourth frame is shown in the composite image here.

Figure 40: Photograph: Martin Kraft / photo.martinkraft.com - Attribution-ShareAlike 3.0 Unported / CC BY-SA 3.0 - http://creativecommons.org/licenses/by-sa/3.0/legalcode

Figure 54: Photographs © 2007 Daniel Treat. Used by permission.

Figure 68: Creative Commons Attribution Share-Alike 3.0 Unported. Author JJ Harrison

Figure 69: This file is licensed under the Creative Commons Attribution 2.0 Generic license. Author: Chris Phutully

Figure 70: Photograph by Bitbanger Labs / www.thepixelstick.com. Reproduced by permission

Figure 74: Images © 2009 K-Sounds. Reproduced by permission. Visit K-Sounds online at ksounds.com.

Figure 122: This image is licensed under the GNU Free Documenation License, Version 1.2 only. "Bombeta de Llum.JPG". Color temperature adjusted cooler for inclusion in this book.

Figure 124: This file is licensed under the Creative Commons Attribution-Share Alike 3.0 Unported license. Author: Alex:D

Figure 135: This file is licensed under the Creative Commons Attribution-Share Alike 3.0 Unported license. Author: Sun Ladder

Figure 137: This file is licensed under the Creative Commons Attribution 3.0 Unported

Trademarks referenced in this book

All trade marks are properties of their respective owners.

Adobe
American Furniture Warehouse
Apple
Audi
Bitbanger Labs
Bombardier
Cadillac
Canon
Caesars Palace
CityCenter
College America
Comfort Dental
Cosmopolitan of Las Vegas
Denny's
DLP (a trademark of Texas Instruments)
Ed Carroll
EliteBook
Famous Footwear
GM, GMC
Google
Hard Rock Cafe
Hewlett-Packard, HP
High Roller
HTC
Hyundai
IrfanView
Kellys Brew Pub
Kia
Kodak
Korg
Lexus
LG
Louis Vuitton
Macintosh
McDonald's
MGM Grand
Microsoft
Midiman / M-Audio
Nikon
Panasonic
Paris Las Vegas
Photoshop
pixelstick
Planet Hollywood
Rayovac
Saab
Sands
Shell
SKY
Tesla
Venetian
YouTube

Index

W

X

Z

Printed in Great Britain
by Amazon

11733360R00089